Renewable Energy and Energy

Renewable Energy and Energy Efficiency

Assessment of Projects and Policies

Aidan Duffy
Professor
School of Civil and Structural Engineering
Dublin Institute of Technology

Martin Rogers
Senior Lecturer
Dublin Institute of Technology

Lacour Ayompe
Researcher
International Energy Research Centre

WILEY Blackwell

This edition first published 2015
© 2015 by John Wiley & Sons, Ltd

Registered office
John Wiley & Sons, Ltd, The Atrium, Southern Gate, Chichester, West Sussex, PO19 8SQ,
United Kingdom.

Editorial offices:
9600 Garsington Road, Oxford, OX4 2DQ, United Kingdom.
The Atrium, Southern Gate, Chichester, West Sussex, PO19 8SQ, United Kingdom.

For details of our global editorial offices, for customer services and for information about how to
apply for permission to reuse the copyright material in this book please see our website at
www.wiley.com/wiley-blackwell.

The right of the author to be identified as the author of this work has been asserted in
accordance with the UK Copyright, Designs and Patents Act 1988.

Library of Congress Cataloging-in-Publication Data applied for.

ISBN: 9781118631041

A catalogue record for this book is available from the British Library.

Wiley also publishes its books in a variety of electronic formats. Some content that appears in
print may not be available in electronic books.

Cover image: iStockphoto © demachi.

Contents

Symbols, Units and Abbreviations   ix
About the Companion Website   xv

1 Introduction   **1**
   1.1 Background   2
   1.2 Aim   4
   1.3 Aspects of renewable energy project appraisal   6
   1.4 Book layout   8
   References   10

2 Technologies   **11**
   2.1 Introduction   11
   2.2 Key concepts   11
      2.2.1 Heat of combustion   12
      2.2.2 Efficiency   12
      2.2.3 Rated power and energy   12
      2.2.4 Capacity and availability factors   13
      2.2.5 Technology learning   13
   2.3 Electrical power generation   14
      2.3.1 Natural-gas-fired power plant   14
      2.3.2 Coal-fired power plant   15
      2.3.3 Hydropower   17
      2.3.4 Wind power   19
      2.3.5 Ocean energy   22
      2.3.6 Photovoltaics   25
   2.4 Heat generation   28
      2.4.1 Boilers   28
      2.4.2 Solar water heaters   30
   2.5 Combined heat and power   34
      2.5.1 Micro-CHP   36
      2.5.2 CHP engines   37
      2.5.3 CHP turbines   37
      2.5.4 Combined heat, power and cooling   38
   2.6 Energy storage   39
      2.6.1 Electrical   40
      2.6.2 Pumped hydroelectric storage   40
      2.6.3 Compressed air energy storage   42
      2.6.4 Thermal energy storage   44

2.7	Energy efficiency	45
	2.7.1 Thermal insulation	46
	2.7.2 High-efficiency lighting	48
	References	50

3 Modelling Energy Systems — **53**

3.1	Introduction	53
3.2	System, model and simulation	54
	3.2.1 Systems	54
	3.2.2 Models	58
	3.2.3 Simulation	71
3.3	Modelling and simulating energy systems	76
	3.3.1 Steps in simulating energy projects	76
	3.3.2 Simulation tools	79
	3.3.3 Data sources	79
3.4	Case studies	83
	3.4.1 Office PV system	83
	3.4.2 Gas heat pump for data room cooling	87
	3.4.3 Compressed air energy storage	90
3.5	Conclusions	93
	References	95

4 Financial Analysis — **97**

4.1	Introduction	97
4.2	Fundamentals	98
	4.2.1 Investor perspective	98
	4.2.2 Types of projects and decisions	99
	4.2.3 Cash flows	100
	4.2.4 Real and nominal prices	104
	4.2.5 Present value	106
	4.2.6 Discount rates	109
	4.2.7 Taxation and depreciation	112
	4.2.8 Unequal project lifespan	114
4.3	Financial measures	116
	4.3.1 Payback and discounted payback periods	117
	4.3.2 Return on investment	120
	4.3.3 Profitability index and savings-to-investment ratio	121
	4.3.4 Net present value	123
	4.3.5 Internal Rate of Return	127
	4.3.6 Life cycle cost	131
	4.3.7 Levelised Cost of Energy	132
	4.3.8 Uncertainty and risk	134
	4.3.9 Financial measures compared	136

4.4   Case studies   139
       4.4.1   Municipal bus fleet conversion to compressed
                natural gas   139
       4.4.2   New wind farm development   142
4.5   Conclusion   148
References   149

5   Multi-Criteria Analysis   151
5.1   General   151
5.2   Simple non-compensatory methods   152
       5.2.1   Introduction   152
       5.2.2   Dominance   153
       5.2.3   Satisficing methods   155
       5.2.4   Sequential elimination methods   157
       5.2.5   Attitude-oriented methods   158
5.3   Simple additive weighting method   160
       5.3.1   Basic simple additive weighting method   160
       5.3.2   Sensitivity analysis of baseline SAW results   163
       5.3.3   Assigning weights to the decision criteria   164
5.4   Analytic hierarchy process   168
       5.4.1   Introduction   168
       5.4.2   Hierarchies   169
       5.4.3   Establishing priorities within hierarchies   169
       5.4.4   Establishing and calculating priorities   171
       5.4.5   Deriving priorities using an approximation method   172
       5.4.6   Deriving exact priorities using the iterative
                Eigenvector method   173
5.5   Concordance analysis   181
       5.5.1   Introduction   181
       5.5.2   PROMETHEE I   184
       5.5.3   ELECTRE TRI   188
5.6   Site selection for wind farms – a case study from
       Cavan (Ireland)   189
       5.6.1   Introduction   189
       5.6.2   National and international guidance   189
       5.6.3   Decision framework chosen   194
       5.6.4   Decision model utilised to categorise each of the
                18 sites   195
       5.6.5   Selection of potentially suitable sites   198
       5.6.6   Concluding comment on case studies   198
5.7   Concluding comments on MCDA models   200
References   202

6 Policy Aspects   **203**

  6.1   Energy policy context   203

  6.2   Energy policy overview   206

       6.2.1   Policy instruments and targets   206

       6.2.2   Designing policy instruments   208

  6.3   Marginal abatement cost   210

       6.3.1   Environmental life cycle assessment   211

       6.3.2   Estimating marginal abatement costs   221

  6.4   Subsidy design   224

       6.4.1   Types of energy subsidies   224

       6.4.2   Calculating feed-in-tariffs   226

  6.5   Social cost–benefit analysis   230

       6.5.1   Define the objective and identify base case   231

       6.5.2   Identify costs and benefits   231

       6.5.3   Value costs and benefits   233

       6.5.4   Discount the costs and benefits   235

       6.5.5   Interpret results   237

       6.5.6   Assess who bears the costs and benefits   237

       6.5.7   Uncertainty   238

       6.5.8   Make decision   238

  6.6   Case studies   238

       6.6.1   Marginal abatement costs of emission mitigation
              options in a building estate   238

       6.6.2   PV feed-in-tariff design   243

  6.7   Conclusions   248

    References   248

Appendix A: Table of Discount Factors   251

Index   253

Symbols, Units and Abbreviations

Abbreviations

AC	Alternating Current
AHP	Analytic Hierarchy Process
BAU	Business as Usual
BAWT	Building Augmented Wind Turbine
bbl	Barrel of oil
BOS	Balance of System
CAES	Compressed Air Energy Storage
CAPEX	Capital expenditure
CBA	Cost-benefit Analysis
CCGT	Combined Cycle Gas Turbine
CCS	Carbon Capture and Storage
CF	Capacity Factor
CHP	Combined Heat and Power
CHPC	Combined Heat and Power and Cooling
CNG	Compressed Natural Gas
CPC	Compound Parabolic Collector
CPI	Consumer Price Index
DC	Direct Current
EDC	Engine-driven Chiller
EIA	Environmental Impact Assessment
ETC	Evacuated Tube Collectors (SWHS)
ETS	Emissions Trading Scheme
FIT	Feed-in Tariff
FPC	Flat Plate Collector (SWHS)
GFA	Gross Floor Area
GHG	Greenhouse Gas
GHP	Gas Heat Pump
GWP	Global Warming Potential
HAWT	Horizontal-axis Wind Turbine
HHV	Higher (gross) heating value
HICP	Harmonised Index of Consumer Prices
HPS	High-pressure Sodium (lamp)
HVAC	Heating, Ventilation and Air Conditioning

IHA	International Hydropower Association
I-O	Input-output (LCA)
IRR	Internal Rate of Return
LCA	Life Cycle Assessment
LCC	Life Cycle Cost
LCE	Life Cycle Emissions
LCOE	Levelised Cost of Energy
LED	Light Emitting Diode
LHS	Latent Heat Storage
LHV	Lower (net) heating value
LPG	Liquid Petroleum Gas
MAC	Marginal Abatement Costs
MARR	Minimum Acceptable Rate of Return
MAUT	Multi-attribute Utility Theory
MCDA	Multi-Criteria Decision Analysis
MIRR	Modified Internal Rate of Return
NHA	National Heritage Area
NPV	Net Present Value
O&M	Operation and Maintenance
OCGT	Open Cycle Gas Turbine
PCM	Phase Change Material
PEM	Proton Exchange Membrane (fuel cell)
PHS	Pumped Hydroelectric Storage
PM10	Particulate Matter (<10μm)
PP	(Simple) Payback Period
PPA	Power Purchase Agreement
PSH	Peak Sun Hour
PV	Photovoltaic
ROC	Renewable Obligation Certificate
ROCE	Return on Capital Employed
RoI	Return on Investment
SAC	Special Area of Conservation
SAW	Simple Additive Weighting
SEA	Strategic Environmental Assessment
SHS	Sensible Heat Storage
SMP	System Marginal Price
SPF	Shadow Price Factors
SWHS	Solar Water Heating System
TES	Thermal Energy Storage
TUoS	Transmission Use of System
TYM	Typical Meteorological Year
VAWT	Vertical-axis Wind Turbine
VSD	Variable Speed Drive
WECS	Wind energy conversion system

Symbols and Units

A	Area	m^2
A	Annuity Factor (Chapter 6)	dimensionless
C	Cost	€
CBR	Cost-benefit Ratio	dimensionless
CDF	Cumulative Discount Factor	dimensionless
CF	Capacity Factor	dimensionless
CF	Net Cash Flow	€
$CO_2\text{-}eq$	Carbon dioxide equivalent	g
COP	Coefficient of Performance	dimensionless
C_p	Power Coefficient (wind turbine)	dimensionless
C_p	Specific Heat Capacity	J/kg °C
CPI	Consumer Price Index	dimensionless
CS	Capital Subsidy	€/W
D	Debt	€
d	Discount Rate	%
DF	Discount Factor	dimensionless
DPP	Discounted Payback Period	y
E	Equity	€
E	Energy (or Electrical Energy)	J or Wh
e	Inflation	%
EAC	Equivalent Annual Cost	€/y
EI	Emissions Intensity	g CO$_2$-eq/€
F	Cash Flow	€/time interval
FIT	Feed-in Tariff	€/Wh
g	Acceleration due to gravity	m/s^2
G_t	In-plane Solar Radiation	W/m^2
H_{m0}	Significant Wave Height	m
HR	Heat Rate	kJ/kWh
irr	Internal Rate of Return	%
LCC	Life Cycle Cost	€
LCE	Life Cycle Emissions	gCO$_2$-eq
$LCOE$	Levelised Cost of Energy	€/Wh
LR	Learning Rate	%
M	Mass	g
$\dot{m}$	Fluid mass flow rate	kg/s
MAC	Marginal Abatement Costs	€/gCO$_2$-eq
MAD	Mean Absolute Deviation	dimensionless
$MAPE$	Mean Absolute Percentage Error	dimensionless
$MARR$	Minimum Acceptable Rate of Return	%
$mirr$	Modified Internal Rate of Return	%
MPE	Mean Percentage Error	dimensionless
N	Number	dimensionless

NPV	Net Present Value	€
P	*Power*	W
P	Cost	€
PI	Profitability Index	dimensionless
PP	(Simple) Payback Period	y
PR	Progress Ratio	dimensionless
Q	Fuel	Wh
Q	Heat	Wh
Q	Quantity	g, l, m^3 , Wh, etc
r	Return (financial)	%
ROCE	Return on Capital Employed	%
RoI	Return on Investment	%
SF	Solar Fraction	dimensionless
SIR	Savings-to-investment Ratio	dimensionless
t	Time	y, h, s
T	Tariff	€/Wh
T	Corporate Tax Rate	%
Ta	Tariff	€/Wh
U	Unit Heat Loss Rate (U-Value)	W/m^{2}K
v	Velocity	m/s
WACC	Weighted Average Cost of Capital	%
η	Efficiency	%
ρ	Density	g/m^3
n_p	Payback Period	yrs

Subscript Symbols

aux	Auxiliary
av	Avoided
c	Investment, Capital
comp	Compressor
cw	Chilled Water
d	Debt
dem	Demand
dt	Displaced Technology
e	Equity
el	Electrical
ER	Round-trip
ex	Export
f	Fluid, Fuel
fv	Future value
g	Gas
gen	Generator

h	Heat
i, in	Input, Inflows
i,j,n	year
inv	Inverter
loss	Losses
main	Maintenance
n	Nominal
n	Net
no	Net Operating
o	Output, Outflow
out	Output
pv	Present Value
r	Real
s	Sector
s	Saving
sto	Stored
th	Thermal
TUoS	Transmission Use of System
u	Useful

About the Companion Website

This book's companion website www.wiley.com/go/duffy/renewable provides you with case study material to further your understanding of Renewable Energy and Energy Efficiency.

1 Introduction

Energy-efficient projects use alternative technologies, fuels and management systems to reduce heat and electricity consumption. Renewable energy-supply projects produce heat and electricity using sources of energy which are regenerated over short time periods. Their recent rise to prominence in modern society has been driven by their low environmental impacts relative to fossil-fuelled alternatives. However, as they mature, energy-efficient and renewable energy technologies must demonstrate not only their environmental benefits but also their economic competitiveness. This book focuses on the assessment of projects using approaches that take into account the unique economic, environmental and energy characteristics of renewable and energy-efficient technologies.

The global demand for energy-supply and efficiency projects has never been greater. Between 2012 and 2035, the demand for primary energy and electricity is estimated to increase by half and 70%, respectively, mainly in developing countries, while in developed countries the ongoing shift to energy-efficient and low carbon supply technologies are projected to continue. These trends are driven by many – mostly inescapable – factors: a growing global population, increasing wealth, uncertainty of fossil fuel price, security of supply concerns and enhanced policies to combat greenhouse gas (GHG) emissions and global warming. For example, by 2013, China, the European Union (EU) and Japan had adopted emission-reduction targets, while California, Australia, New Zealand and the EU had introduced carbon emissions trading schemes. Assuming the implementation of such existing policy commitments only, it is projected that between 2010 and 2035, a $37tn investment will be required in the world's energy-supply infrastructure and as much as $11.8tn will be spent on energy-efficient measures across all economic sectors (IEA, 2012).

Renewable Energy and Energy Efficiency: Assessment of Projects and Policies, First Edition.
Aidan Duffy, Martin Rogers and Lacour Ayompe.
© 2015 John Wiley & Sons, Ltd. Published 2015 by John Wiley & Sons, Ltd.
Companion Website: www.wiley.com/go/duffy/renewable

Each of the myriad of energy efficiency and supply projects which will comprise these investments must be identified, shortlisted, modelled and economically assessed before it can be financed and implemented. Some will be very large investments such as nuclear or hydro power schemes; others will be small energy-efficient measures such as the installation of domestic attic insulation. All require a systematic approach to assessing their relative costs and benefits. The intention of this book is to present and illustrate the assessment tools necessary to make these decisions as efficiently as possible.

1.1   Background

The history of assessing the costs and benefits of energy projects is probably as long as humans have been harnessing energy for their needs. Hunter-gatherers must have recognised that the advantages of cooking, light and warmth from fires outweighed the time and effort involved in collecting the necessary fuel. However, it was not until the 18th century that the formal process of investment appraisal (or capital budgeting) emerged as a discipline, which focused on quantifying the benefits of long-term capital investments to companies. Assessing the cost-effectiveness of energy investments became much more important as a result of the 1973 oil and 1979 energy crises, which resulted in real oil prices increasing from a long-term historic average of about $20/barrel ($/bbl) to $60 and then over 100$/bbl (Figure 1.1). This heralded a much greater level of interest in energy-efficient and renewable

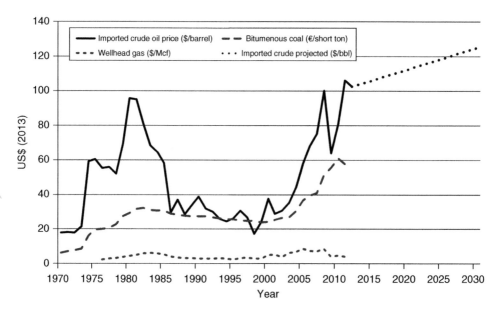

Figure 1.1   Real prices ($2013) of US imported crude oil, wellhead natural gas and bituminous coal, 1970–2012 (US EIA, 2013), and projected oil price in 2030 (IEA, 2012).

energy-supply technologies as economic alternatives to fossil fuels. With the long-term rise in fossil fuel prices since the 1970s and projections for this trend to continue, the investment appraisal of energy projects has continued to become increasingly important (see Figure 1.1).

While the process of investment appraisal has been developed to meet the needs of the private investors of a project, the costs and benefits to the wider community are often ignored, often because they have no market value and are thus difficult to quantify. As energy-supply infrastructure became more widely deployed in developed countries in the mid-20th century and as societies became more environmentally and socially aware, these costs and benefits became more apparent. For example, coal combustion for industrial and domestic heating caused smog, resulting in increased morbidity, higher health costs and lost productivity; hydroelectric dams were built without sufficient consideration of their undesirable impacts on agriculture, fishery and local communities. As these impacts often have no direct market value and are, therefore, difficult to monetise, methods other than investment appraisal become necessary. One solution to this was cost–benefit analysis (CBA or benefit–cost analysis), which was first developed in the mid-19th century but was not used in practice until the 1930s for assessing the attractiveness to society of large infrastructural projects. CBA is typically used to monetise and compare the costs and benefits of large projects or policies that have societal impacts. It attempts to approximate and account for the monetary values of non-marketed goods and services such as air and water quality, employment impacts or displaced local industry. A project is beneficial where its societal benefits outweigh its costs.

However, many large energy projects are complex and have important attributes that are difficult to either quantify or monetise or both. For example, the visual impact of wind turbines on the landscape may affect house prices for the local population and amenity value for tourists: these effects can be difficult to quantify and value. In projects of public importance where environmental and social criteria assume significant importance, purely economic approaches such as CBA or investment appraisal cannot represent all of the attributes which must be considered for an accurate assessment. The emergence of multi-criteria decision analysis (MCDA) in the 1960s and 1970s attempted to address this failure by allowing impacts on different scales to be compared. It breaks the assessment problem in smaller parts to facilitate analysis and aggregates these in a way that allows a project ranking to be made. MCDA is now widely used to shortlist options for large energy projects of public importance such as hydroelectric dams, transmission infrastructure and wind farms.

Energy policies as well as large, strategically important energy projects, which are supported by the state, must be measured not only by their value for money to their investors and wider society but also by their ability to achieve important national objectives such as GHG emission-reduction targets. In February 2005, the Kyoto agreement came into force obliging many developed countries to limit emissions of GHGs. Since then, emission mitigation has become a key energy policy objective for many industrialised countries. In order to develop and monitor policies, project appraisal techniques have been extended to measure the societal cost of climate change mitigation. One widely used metric is the marginal abatement cost (MAC) of a technology, which expresses the cost to an economy of reducing emissions by one unit by switching to a more energy-efficient or renewable energy technology. Initially, only operational emissions were considered when estimating MACs, but the low operational and relatively high production and installation emissions of renewable energy systems led to concerns about the accuracy of this approach. This has led to the increasing adoption of life cycle assessment as a method for estimating the whole life (or "cradle-to-grave") emissions of energy systems.

The development of the aforementioned economic and non-economic assessment techniques (investment appraisal, CBA and MCDA) over the past two centuries has been critical to the effective assessment of energy projects. Two more recent developments are the widespread development of personal computers (PCs) and the collection of large energy-related datasets, many of which are in the public domain. These provide both a wide variety of input data and the necessary processing power for energy-related models. PCs now support powerful programming and data analysis packages that can be configured to simulate a wide range of detailed energy systems. Hourly wholesale electricity data are publically available in Ireland; dynamic wind farm output data are freely available in Denmark, and the advent of smart metering leads to collection of energy demand data at the level of individual buildings. These enable the development of detailed dynamic numerical models which are representative of energy conversion and conservation processes, which, until even a decade ago, were not possible. The resulting ability to model accurate cash flows, pollutant emissions and other outputs provides much more knowledge for decision-making purposes than was possible heretofore.

1.2   Aim

The aim of this book is to provide the reader with the tools to shortlist and economically evaluate energy projects as well as gain an appreciation for aspects of energy policy design. Specifically, students will learn

- approaches to the dynamic modelling of energy inputs and outputs to and from a wide variety of projects employing a range of different renewable and energy-efficient technologies;
- how to extend these models to estimate cash flows and GHG emissions;
- ways of parameterising these results in order to quantify the financial and environmental performances of the projects;

- techniques for assessing and shortlisting complex projects involving non-economic impacts; and
- simple methods for designing price supports for energy technologies.

The book is written for students and practitioners alike. Undergraduate and postgraduate students will be introduced to the economic performances of a variety of technologies and to the basic concepts and frameworks of project financial assessment. Facilities' manager will learn how to provide evidence for business plans outlining their proposals for the energy retrofit and upgrade of their assets. Design engineers will benefit from understanding how to economically optimise their design solutions rather than sizing the plant and equipment to meet peak loads. Planners of large infrastructural projects will be introduced to systematic techniques for site and project screening before undertaking a detailed economic appraisal of a smaller set of project alternatives. Policy makers and planners will be introduced to the fundamentals of subsidy design for renewable and energy-efficient technologies.

The book deals primarily with renewable energy-supply and energy-efficient technologies. Renewable energy supply relates to energy conversion technologies, which use sources of energy that are naturally regenerated on a short (human) time scale, such as solar, wind, ocean, biofuels and geothermal energies. Typical technologies that convert renewable energy sources into electricity include wind turbines, solar photovoltaics (PV), biomass-driven gas and steam turbines, geothermally driven steam turbines, concentrating solar power, tidal barrages and a variety of wave-powered devices. Thermal conversion technologies are very diverse and include solar water heaters, biomass and biogas boilers and stoves and geothermal technologies. Non-renewable energy supply relates primarily to fossil fuels such as oil, coal and natural gas, which cannot be created in a human time scale and typically take many millions of years to form. Fossil-fuelled electricity generation usually employs various gas- and steam-turbine technologies, but also includes reciprocating engines, while heat generation is normally undertaken using boiler technologies and, to a lesser extent, stoves. We do not deal specifically with nuclear power in this book because this is a large topic in itself that, at the time of writing, has an uncertain medium-term future for political, economic and environmental reasons. Nonetheless, many of the principles described here can be directly applied to this technology.

Energy-efficient projects involve the use of alternative technologies, fuels and management systems to deliver the same level of service or output using less energy (irrespective of the energy-supply source). Therefore, any technology, fuel or management system has the potential to be energy efficient, because this classification is gained by comparing it to the displaced alternative. An obvious example is increasing the amount of insulation in a building to reduce heat losses, so that the same level of thermal comfort can be provided using less heat and, therefore, fuel. Burning gas instead of coal in a thermal power station can result in the consumption of less primary energy and, in this situation, may be viewed as an energy-efficient technology. It should be highlighted that although fossil-fuelled technologies are not the focus of this book, the concepts presented are equally valid for their assessment.

The learning approach adopted involves explaining each theory and providing an example in close proximity in order to illustrate and embed the concept. Larger case studies are also included to demonstrate the combination of different concepts for more complex examples. We attempt to give examples for a wide variety of industry sectors and applications to make the book as broadly relevant as possible. Examples are given for the domestic, commercial, energy and industrial sectors using single and multiple electrical and thermal technologies such as PV, solar water heaters, wind turbines, wave power, combined heat and power, boilers and insulation. In addition, a number of policy examples that illustrate feed-in-tariff and capital subsidy design are given. We attempt to make this book as practical as possible, so that the reader is able to easily apply the concepts to projects of personal interest. For this reason, the examples and case studies are made available online (www.wiley.com/go/duffy/renewable), which help the reader to gain a detailed understanding of the techniques used and apply them directly to problems of personal interest.

Finally, this book adopts a bottom-up 'engineering' approach to the financial appraisal of renewable energy projects. This involves the modelling, simulation and economic parameterisation of individual energy projects in isolation to the market in which they operate.

1.3   Aspects of renewable energy project appraisal

In general, the appraisal of renewable energy and energy-efficient projects is no different to the assessment of any other capital projects. Although we will see that some project performance measures are specific to the field, the main appraisal techniques described here such as investment appraisal, CBA, and multi-criteria analysis are widely applied to other investments, both large and small. Nevertheless, renewable and energy-efficient projects do have unique characteristics, which the assessor must be aware of in order to undertake a proper assessment.

Many renewable and energy-efficient projects are characterised by high initial investment costs and low operational costs. This is true for technologies such as wind, PV and solar thermal as well as energy-efficient measures such as insulation. Conventional fossil-fuelled plant, on the other hand, has lower capital costs as a proportion of total life cycle costs with relatively higher operational outgoings because of the ongoing need to purchase fossil fuels. This means that renewable energy and energy-efficient supply projects are generally less exposed to fluctuations in variable costs as compared to fossil-fuelled ones due to the high price volatility of fuel inputs, particularly oil and its derivatives. Renewables do remain exposed to fluctuations in revenues resulting from changes in the unit cost of energy outputs, such as electricity and heat, as does conventional plant, although input and output prices tend to move together, thus acting as a natural 'hedge' to revenue risk for fossil-fuelled plant. Often, the 'revenues' in renewable or energy-efficient projects are avoided costs such as the cost of grid electricity displaced by embedded generators. Revenues from many renewable projects may also include long-term price supports such as feed-in-tariffs, which provide a fixed production tariff or tariff floor. However, these can represent a significant risk because a single regulatory decision can greatly alter the basis of an initial investment decision. This political risk is exacerbated by the long payback periods needed for many renewable energy technologies. For example, PV feed-in-tariffs were reduced in the United Kingdom, Spain, Germany and Bulgaria, between 2009 and 2011. Societal imperatives can also shift quickly: the Great Recession of 2009 focused public debate on economic growth and employment while costly emissions' mitigation policies dropped down the priority list. The identification and quantification of project risk are, therefore, an important task in many renewable and energy-efficient project assessments.

The fast-changing energy and renewables landscape results in other risks too. Technology costs are evolving quickly: real capital costs of installed US commercial PV system have more than halved in the 15 years between 1998 and 2013 (Feldman et al., 2012), whereas the development of hydraulic fracturing technology has been associated with a drop in nominal US wellhead natural gas prices from \$6.25 to \$2.66/1000 ft^3 between 2007 and 2012 (US EIA, 2014). Therefore, the timing of investments in renewable energy and energy-efficient projects and policies is particularly important. For example, investing under conditions of strong global growth is likely to be more attractive as energy prices are likely to be higher giving greater certainty to short- and medium-term revenues. Moreover, technology costs in the future are likely to be lower, possibly resulting in better returns to the private investor and lower technology subsidies.

Many renewable energy technologies rely on national subsidies for a variety of reasons, not least because they may not be competitive with conventional alternatives. The approach is controversial as governments do not have a reputation for 'picking winners', particularly in a field as technologically complex as energy conversion, storage, transmission and efficiency. These subsidies include feed-in-tariffs, capital subsidies, tax rebates and renewable obligations certificates. Opponents argue that putting a price on the negative effects of fossil fuels using a carbon tax is a more efficient approach because the market would adopt the technology with the lowest marginal abatement cost, thus resulting in lower overall societal costs as compared with subsidies. However, renewables' subsidies are regarded by others as important in encouraging investment in emerging low carbon technologies, accelerating market growth and reducing technology costs. State investments in onshore wind since the 1990s, for example, have greatly contributed to a decrease of about two-thirds in the real cost of wind power plant over the past two decades. Indeed, the costs of renewable technologies are falling more rapidly than conventional technologies because they are typically less mature. However, while policy supports can result in widespread technological deployment, learning and cost reductions, they can also result in supply constraints and increased market prices. Policy makers must apply project appraisal techniques to answer the questions: What is the minimum support necessary to support a technology? Is this cost-effective in supporting key government policies such as emissions mitigation? It is important that where subsidies are introduced they represent value for money for the taxpayer.

Project assessors should be aware that renewable energy-supply technologies do not always offer identical outputs to the conventional alternatives. A unit of electricity from a wind turbine is not the same as that from a thermal power station because the latter is almost always available when it is needed (i.e. it is 'dispatchable'), whereas the former is only available when the wind is blowing and its availability cannot be guaranteed when needed (and it, therefore, is 'non-dispatchable'). An accurate comparative analysis should always compare like-with-like; for example, storage and backup should be included with intermittent renewable generation when comparing it with dispatchable plant, so that identical levels of service are provided in each case. This approach should be considered when comparing any intermittent technology (wind, solar and ocean). However, when compared to conventional alternatives, renewable energy projects can provide additional benefits to society over fossil-fuelled alternatives, which should be considered as part of the assessment process. These include emissions reductions, local employment as well as increased national security of energy-supply due to reduced import dependency (in net energy importing countries only). Social costs imposed by renewable and energy-efficient projects should also be included.

1.4   Book layout

There are five main chapters in this book that introduce the reader to the techno-economic characteristics of renewable and energy-efficient systems, financial and non-financial project assessment methods and aspects of energy policy. Each chapter includes an initial content overview before describing relevant theory; short examples are provided throughout, which apply this theory to practical applications of renewable energy and energy-efficient projects. Chapters 3–6 include concluding comments, which highlight the key concepts introduced. Case studies are included at the ends of chapters, which illustrate how complete renewable energy projects might be assessed using the main concepts introduced.

Chapter 2, 'Technologies', describes a variety of renewable energy and energy-efficient technologies, which are necessary to understand the examples and case studies described in the book. The descriptions mainly focus on those aspects that are necessary for subsequent modelling and appraisal such as efficiencies and other operational parameters, investment costs, operating and maintenance costs and environmental emissions. This is by no means a comprehensive overview of all relevant technologies because this is not the focus of the book.

The foundation of almost all financial measures of energy project performance is an accurate cash flow. For energy projects, all cash flows are directly related to energy flows to and from the system being considered. For example, the cost of running a gas-fired boiler is related to how efficient it is at converting the gas input into the necessary heat output. Quantities of gas used and heat produced represent the main costs and benefits of the system and, together with capital cost, largely determine its financial performance. Similarly, environmental impacts such as GHG emissions are largely determined by the gas inputs to the system. Therefore, Chapter 3, 'Modelling Energy Systems', is dedicated to system definition, modelling and simulation.

Chapter 4, 'Financial Analysis', uses these cash flows to create financial measures – or parameters – for renewable energy and energy-efficient projects. First, fundamental concepts are introduced, which are necessary for converting project cash flows into useful parameters. A wide variety of parameters are then presented and their strengths and weaknesses in different contexts discussed. Those of particular relevance to assessing renewable energy projects are highlighted.

Not all projects can be compared on purely economic grounds. Many other advantages and disadvantages of a particular project option may be important. For example, social, political and environmental dimensions may be particularly important for large infrastructural projects such as the construction of hydroelectric dams or the routing of large overhead transmission lines. Chapter 5, 'Multi-criteria Analysis', offers alternative methods for shortlisting and selecting projects using MCDA techniques.

Chapter 6, 'Policy Aspects', combines these financial techniques with environmental assessment methods and extends them to introduce basic concepts in policy design. An initial review of policy options for emission mitigation is followed by an overview of life cycle assessment and methods for quantifying GHG emissions from different renewable energy and energy-efficient projects. The chapter explains marginal abatement costs and subsidy design and gives a short introduction to social CBA.

Case studies are provided at the end of Chapters 3–6, which demonstrate the application of many of the key concepts introduced in these chapters. Case studies include energy and cash flow models (commercial PV systems, gas heat pumps for data room cooling, compressed air energy storage), financial appraisals (converting a bus fleet to compressed natural gas fuel, wind farm appraisal), non-economic analysis (wind farm site selection) and policy-related assessments (MAC estimation and domestic PV feed-in-tariff design). Case study spreadsheet calculations can be accessed at www.wiley.com/go/duffy/renewable.

References

Feldman, D., Barbose, G., Margolis, R., Wiser, R., Darghouth, N. and Goodrich, A. (2012) Photovoltaic (PV) Pricing Trends: Historical, Recent, and Near-Term Projections, US Department of the Environment. [online]. Available at www.nrel.gov/docs/fy13osti/56776.pdf. Accessed 20 Oct 2014.

IEA (2012) *World Energy Outlook 2012*. Organisation for Economic Co-operation and Development (OECD)/International Energy Agency (IEA).

US EIA (2013) *Short-Term Energy Outlook Real and Nominal Prices*, US Energy Information Administration [online]. Available at www.eia.gov/forecasts/steo/realprices/; www.eia.gov/coal/data.cfm#prices. Accessed 20 Oct 2014.

US EIA (2014) Natural Gas Price Data, US Energy Information Administration [online] http://www.eia.gov/dnav/ng/ng_pri_sum_dcu_nus_a.htm. Accessed 30 Oct 2014.

2 Technologies

2.1 Introduction

This chapter provides information on the technical and economic characteristics of renewable energy and energy efficient systems that is necessary for the energy modelling, financial appraisal, policy analysis and non-economic assessment methods described in Chapters 3–6. It covers a variety of technologies involved in energy generation, use and storage, many of which are used in examples and case studies throughout the book. The technologies are grouped into broad areas, which include electrical power generation; heat generation; combined heat, power and cooling; energy storage and energy efficiency. Fossil-fuelled technologies are included because although they are not renewable, switching to more efficient fossil-fuelled plant is an important source of energy efficiency gains in many situations.

For each of the energy systems covered, a description of the technology is presented followed by an analysis of its energy and economic characteristics. Emphasis is placed on parameters relevant to the appraisal of these technologies which include energy conversion processes, efficiencies, capital costs and operation and maintenance (O&M) costs. Different technologies are at different levels of maturity, and life cycle costs will decrease more rapidly for some as compared to others as they become more widely deployed. Therefore, where possible, technology learning rates are also presented.

2.2 Key concepts

A number of concepts which are common to many of the technologies are discussed in this chapter. These primarily focus on fuel characteristics, energy efficiency, power and energy, measures of plant availability and technology learning.

2.2.1   Heat of combustion

Almost all fossil-fuelled and biomass energy conversion technologies involve combustion. In the case of boilers, this thermal energy is used directly for a thermal end use. In electricity production, it is normally converted to rotational mechanical energy and then to electricity using an alternator. The heat of combustion of the fuel (or the 'heating value' or 'calorific value') is an important parameter in estimating the final thermal or electrical energy output. The heating value of any fuel is the energy released per unit mass when the fuel is completely burned. It depends on the chemical characteristics of the fuel and on the state of water molecules in the final combustion products. The fuel energy input can be evaluated as either the higher (gross) heating value (HHV) or the lower (net) heating value (LHV) of the fuel. HHV refers to a condition under which the water condenses out of the combustion products. As a result of this condensation, both latent and sensible heats affect the heating value. LHV, on the other hand, refers to the condition under which water in the final combustion products remains as vapour (or steam). The steam does not condense into liquid water; therefore, the latent heat is not accounted for. It is, therefore, important to use the same heating value when comparing the efficiency of different energy conversion heating systems.

2.2.2   Efficiency

The energy efficiency of an energy conversion device is defined as the useful energy output divided by energy input. The amount of energy inputted into a system is typically less (but never greater) than its energy output due to losses. These occur for many reasons including frictional, thermal and electrical resistance.

It is important to consider the system you are analysing when estimating its energy efficiency. For example, in order to estimate the amount of fuel consumed in providing electrical lighting in a home, it is important to consider all steps in the process: electricity production by the power plant, transmission and distribution losses, light bulb efficiency and any shading effects. In this case, the overall efficiency could be lower than 25%. However, if the efficiency of the electricity produced by the power plant only is considered then the efficiency would be in the range of 40–60%.

2.2.3   Rated power and energy

The size of a power plant or energy conversion technology is normally discussed in terms of its rated power. This is the maximum electrical or thermal power that it is capable of producing under normal operating conditions, which is recommended by the manufacturer. Rated power is typically given in kilowatts (kW) or megawatts (MW). It is analogous to brake horsepower in a car. Domestic energy conversion (e.g. domestic boilers) devices are normally measured in kilowatts, while power stations are normally measured in megawatts. Devices normally operate at a power output between a minimum threshold and the maximum rated power.

The energy output from an energy conversion technology is the product of the power output and the time of operation at this output. Although the international unit for energy is joule (J), we will normally measure energy in kilowatt-hours (kWh), megawatt-hours (MWh) or gigawatt-hours (GWh), where 1 Wh is equivalent to 3600 J.

Energy cannot be 'consumed' because it can only be converted from one form to another. We, therefore, attempt to refer to energy 'use' rather than 'consumption'; fuels, however, can be consumed.

2.2.4 Capacity and availability factors

The capacity factor (CF) or load factor of an energy conversion technology is its energy output as a fraction of the energy that could be generated at its rated power over the same period of time. CF is often expressed in equivalent full-load hours representing the time over which the technology would have to operate annually at its rated power to reach the measured annual generation, for example, a CF of 30% corresponds to an equivalent full-load use of 2628 h/annum.

Plant availability factor is the percentage of time that it is available for energy conversion. *Planned outages* are scheduled periods when the plant operation is stopped for anticipated maintenance. *Forced outages* are unscheduled maintenance periods when the plant unexpectedly ceases to operate normally and requires maintenance. It is important to consider plant availability in any financial analysis.

2.2.5 Technology learning

'Technology learning' or 'learning-by-doing' occurs as the market for a technology develops. It results in lower technology costs for a wide variety of reasons, including technological advances, process optimisation, manufacturing economies of scale and supply chain development. The rate of technology learning has been rapid for a number of renewable technologies such as wind and photovoltaics (PV). It is important that this phenomenon be considered when designing policy supports or making investment decisions for large projects with long lead-in periods.

Learning curves are expressed as

$$C(x_t) = C(x_0)\left(\frac{x_t}{x_0}\right)^b \tag{2.1}$$

where $C(x_t)$ is the cost in year t, $C(x_0)$ is the cost in an arbitrary starting year, b is the learning parameter or learning elasticity parameter or rate of innovation, x_t is the cumulative installed capacity in year t and x_0 is the cumulative installed capacity in the starting year.

With every doubling of cumulative production, costs decrease to a value expressed as the initial cost multiplied by a factor called the 'progress ratio'. The progress ratio is expressed as

$$PR = 1 - LR = 2^b = 2^{\ln \frac{C(x_t)}{C(x_0)} / \ln \frac{x_t}{x_0}} \tag{2.2}$$

where PR is the progress ratio and LR is the learning rate.

Learning rates for industrial products including energy efficient and supply technologies typically range from 10% to 30%.

2.3 Electrical power generation

2.3.1 Natural-gas-fired power plant

Natural-gas-fired power plants have recently become more popular for electricity generation because they can be more efficient compared to other fossil-fuelled technologies, generate fewer pollutants and due to the recent fall in North American gas prices. Technologies that use natural gas to generate electricity include steam units, gas turbines and combined cycle units. In steam-generating units, natural gas is burned in a boiler to heat water to produce steam, which is used to turn a turbine and generate electricity. Steam units have low efficiencies, typically ranging between 32% and 35%. In gas turbines, hot gases from burning natural gas are used to turn the turbine and generate electricity. Combined cycle units incorporate both a steam unit and a gas turbine as a single unit, as shown in Figure 2.1. Here, natural gas

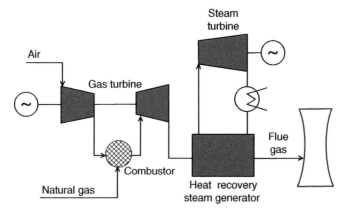

Figure 2.1   Combined cycle gas power plant.

is burned with air and the resulting hot gases generate electricity as in gas turbines, while the waste heat from the gas turbine is used to generate steam, which drives a turbine and generates electricity. Combined cycle plants have efficiencies of 50–60% as a result of the efficient use of the heat generated. Over their lifespan, combined cycle plants are an economically competitive option for new base-load power generation.

The performance of a power plant can be expressed in terms of the heat rate (HR) and thermal efficiency. The electrical energy output from a gas-fired power plant is related to its efficiency and gas fuel input by

$$E_{o,t} = \eta \times Q_{f,t} \tag{2.3}$$

where $E_{o,t}$ is the electricity generated (kWh) over a time period t, $Q_{f,t}$ is the quantity of fuel used (kWh) over a time period t and η is the plant efficiency (%).

The HR is the efficiency of conversion from fuel energy input to electricity output. It is the fuel energy required to generate a unit of electricity (kJ/kWh) and is traditionally used in the industry instead of plant efficiency. HR is related to efficiency by

$$\eta = \frac{3600}{\text{HR}} \times 100 \tag{2.4}$$

The electricity generation cost for natural gas plants is influenced by the number of full operating hours, efficiency, capital (investment) cost, variable and fixed O&M costs. For gas turbine and combined cycle gas turbine (CCGT) plant, capital expenditure includes gas (and steam for CCGT) turbines; balance of plant; engineering, procurement, construction management services and other costs (such as land, inventory capital, spare parts and plant equipment, grid connections, project development, project management). Turbines and balance of plant account for at least half of all investment costs. Variable O&M costs are much higher than fixed O&M costs because the former are dominated by fuel, the most expensive system input. Table 2.1 summarises the characteristics of natural gas power plants.

2.3.2 Coal-fired power plant

Coal-fired power plant generates electricity by burning coal and heating water in a furnace or boiler to produce steam. The steam flows under high pressure through a turbine that turns a generator to produce electricity. The boiler is the main component of a coal-fired power plant, and its optimal design and operation have a large impact on the plant's overall efficiency and emissions. New coal power plant technologies are under constant development, resulting in on-going efficiency gains and reduced emissions. Such technologies include fluidised bed combustion, oxy-fuel combustion, advanced gasification, integrated gasification combined cycle and high performance power systems. Figure 2.2 shows a typical coal-fired power plant.

Table 2.1   Characteristics of selected natural gas power plants.

Parameter	Gas turbine	Combined cycle
Capacity (MW)	211	580
Efficiency (%)	32.8	50.9
Capital cost (€/kW)	360–610	690–1150
Variable O&M costs (€/MWh)	22.50	2.75
Fixed O&M costs (€/(kW yr))	3.95	4.73
Emissions rate (g/MWh)		
SO_2	0.0015	0.0005
NO_x	0.2421	0.0195
PM10	0.0440	0.0155
CO_2	858.2	312.2

Source: U.S. Energy Information Administration, Form EIA-860, 'Annual Electric Generator Report'; NREL (2012).

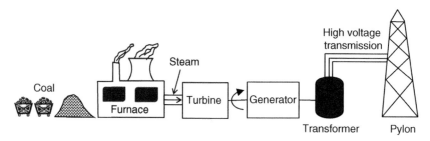

Figure 2.2   Coal-fired power plant.

   The efficiency of a conventional coal-fired power plant using a steam turbine generation system is in the range of 25–40%, and it is predicted that future combined cycle plants will have efficiencies in excess of 60%. Integrated gasification combined cycle is one of the recent concepts of advanced power generation with the most efficient power and lowest emission of pollutants. The energy output characteristics of coal-fired plant are the same as those for gas-fired plant and are given by Equations 2.3 and 2.4.

   The electricity generation costs for coal-fired power plant is influenced by the number of full operating hours, efficiency, capital, variable and fixed O&M costs. For pulverised coal plants, investment costs include turbine equipment; balance of plant/installation (accounting for over 60% of capital costs); engineering, procurement, construction management services and other costs (such as land, inventory capital, spare parts and plant equipment, utility connections, project development, project management). An important international policy objective is the decarbonisation of coal-fired power plant through the use of carbon capture and storage (CCS). Current technologies at demonstration stage can remove 85% of flue gas carbon emissions but the plant efficiency reduces by about 34% (the HR increases from 7.07 to

Table 2.2   Characteristics of selected coal-fired power plant technologies.

Parameter	Pulverised coal	Pulverised coal with carbon capture	Gasification combined cycle	Gasification combined cycle with carbon capture	Flue gas desulphurisation retrofit
Capacity (MW)	606	455	590	520	
Capital cost (€/kW)	1400–2900	2800–7000	1950–4100	5200–7000	270
Efficiency (%)	36.4	27.1	37.8	28.9	
Variable O&M costs (€/MWh)	2.78	4.52	4.91	7.95	2.78
Fixed O&M costs (€/(kW yr))	17.25	26.40	23.33	33.30	17.40
Emissions rate (g/MWh)					
SO_2	0.1405	0.1871	0.1705	0.1935	
NO_x	0.1277	0.1701	0.2230	0.2530	
PM10	0.0281	0.0374	0.0236	0.0268	
Hg (% removal)	90	90	90	90	
CO_2	549.1	108.9	564.0	95.2	

Source: NREL (2012).

10.64 MJ/kWh). Table 2.2 outlines some characteristics of different coal-fired power plant technologies.

2.3.3   Hydropower

Hydropower is a renewable energy resource that uses potential energy stored in water which is converted to kinetic energy when it flows through a duct called the 'penstock' and then into mechanical energy by turning the turbines and, finally, to electrical energy using an electric generator. Hydropower is a mature technology with installations ranging in rated capacity from a few kilowatts electrical (kW(e)) to over 20,000 MW(e). Hydropower normally delivers base-load power but has the potential to be quickly ramped up and down to meet changing electricity demands. Hydroelectric power plants can also be used as black start sources to restore network interconnections and other power stations to operation in the event of a 'blackout', or power outage.

Three main types of hydropower installations are storage schemes, run-of-river schemes and pumped storage. In a storage scheme, a dam is used to store water that is channelled to the turbine, which eventually turns the generator as shown in Figure 2.3. In a run-of-river scheme, the natural flow of a river is channelled through a weir to increase the flow, which is then used to turn a turbine, which in turn turns the generator. A description of pumped hydroelectric storage schemes is found in Section 2.6.2.

The power that can be extracted from flowing water depends on the difference in height between the source and water outflow (head) and

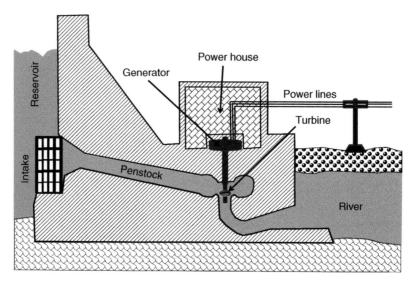

Figure 2.3 Hydropower plant with a storage reservoir.

the volumetric flow rate of the water through the penstock. For a large hydropower dam, the electrical power output is calculated using Equation 2.5 given as

$$P_e = \rho g h Q \eta \tag{2.5}$$

where P_e is the electrical power output (kW), ρ is the density of water (1000 kg/m^3), g is the acceleration due to gravity (9.81 m/s^2), h is the water head (m), Q is the volumetric flow rate (m^3/s) and η is the turbine–generator conversion efficiency (%).

The annual electrical energy output from a hydropower plant is estimated as

$$E_o = P_e \times \text{CF} \times 8760 \tag{2.6}$$

where E_o is the annual electrical energy output (kWh), P_e is the electrical power output (kW), CF is the capacity factor (%) and 8760 is the number of hours in a year.

Hydropower requires relatively high initial investments but typically has the longest lifespan of any generation plant and, in general, low O&M costs. Nevertheless, these investments require regular maintenance including intermittent major scheduled overhauls. Investment costs are highly dependent on location and site conditions. Factors affecting the capital cost and cost of electricity generation include project scale, location, presence and size of storage and use of power generated for base or peak loads or both as well as other benefits derived from flood control, irrigation and freshwater provision.

Average investment costs for large hydropower plants with storage range between 780 and 5750 €/kWh. For small hydropower plants, average investment costs range between 975 and 6000 €/kWh. Annual O&M costs

Table 2.3   Typical characteristics of hydropower projects.

Parameter	Large hydro (≥100 MW)	Small hydro (1–20 MW)	Refurbishment/ upgrade
Installed costs (€/kWh)	780–5750	975–6000	375–750
Annual O&M (% of CAPEX)	2–2.5	1–4	1–6
Capacity factor (%)	25–90	20–95	
LCOE (€c/kWh)	1.50–14.25	1.50–20.25	0.75–3.75
Efficiency (%)	90	90	

LCOE calculations assume a 10% cost of capital.
Source: IRENA (2012a) and OECD/IEA (2010).

range between 2.0% and 2.5% of installed costs (3.75–15.00 €/MWh) for large hydropower plants, while they range between 1% and 4% (7.50–30.00 €/MWh) for small hydropower plants. Annual O&M costs are estimated at between 3.75 and 15.00 €/MWh for large hydro plants, and between 7.50 and 30.00 €/MWh for small hydro plants (OECD/IEA, 2010).

CFs for large hydro schemes typically range from 25% to 90% and from 20% to 95% for small schemes depending on the region and whether they are used as base load or peaking plants. In 2009, the CF for hydropower plants in different continents varied between 32.8% and 56.5% with a global average of 42.6% (IHA, 2012). Hydropower installations are durable and robust with a lifespan between 50 and 100 years. The cost of power generation varies widely, depending on individual project characteristics but typically ranges from 37.5 to 75.0 €/MWh (OECD/IEA, 2010). Table 2.3 details the typical characteristics of hydropower plants.

By the end of 2010, there was an estimated 935 GW of hydropower capacity installed worldwide. In 2011 and 2012, there was an estimated 25–30 GW and 27–30 GW, respectively, of new hydropower commissioned globally, raising the total installed hydropower capacity to 990 GW (IHA, 2010, 2012, 2013; REN21, 2013). Figure 2.4 shows the evolution of global annual consumption of hydroelectric power based on gross primary generation. Between 1965 and 2012, global consumption of hydroelectric power grew by an annual average of 58.5 TWh.

2.3.4   Wind power

Wind energy systems (WESs) use the force of moving air to power wind turbines and generators to produce electricity. These units usually consist of foundations, a tower, rotor (blades, hub and shaft), gear box, generator, control equipment, power conditioning equipment and braking system to protect them from high winds and extreme weather conditions. As the wind does not blow with the same intensity all the time, the power output from wind turbines is intermittent. Wind turbines do not, therefore, replace an equal amount of fossil fuel capacity, but they do replace the quantity of electrical energy produced by other power-generating sources.

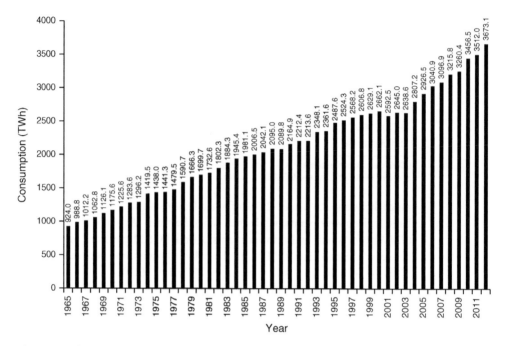

Figure 2.4   **Global annual consumption of hydroelectric power (*Source:* BP, 2013).**

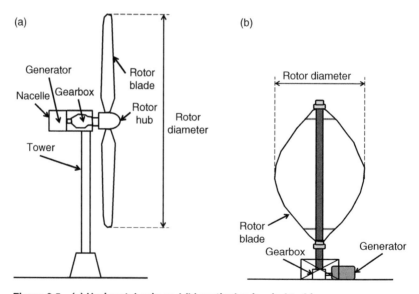

Figure 2.5   **(a) Horizontal-axis and (b) vertical-axis wind turbines.**

Wind turbines are available in three broad categories: horizontal-axis wind turbines (HAWTs), vertical-axis wind turbines (VAWTs) and building augmented wind turbines (BAWTs). HAWTs (shown in Figure 2.5) are commonly used in large-scale wind projects whereas all three categories are used in small-scale applications. VAWTs require less land space as compared to HAWTs. Wind energy projects normally comprise multiple adjacent units know as 'wind farms', which can be situated either onshore or offshore.

The electricity produced by a wind turbine depends on the local wind climate, wind turbine technical specifications, site characteristics and losses. The local wind resource is the most important factor affecting the profitability of wind energy investments and also explains most of the differences in the cost per kilowatt-hour of electricity generated in different countries and projects.

The CF of a wind turbine expresses the percentage of time for which a wind turbine produces electricity in a representative year at full load and largely depends on the site wind resource and turbine characteristics; it may also be affected by network and system constraints (often referred to as 'curtailment') limiting the amount of wind energy which can be produced at a particular time. The CF for onshore wind turbines ranges between 25% and 40%, while it varies between 35% and 45% for offshore wind turbines (REN21, 2013). The downtime (comprising planned and forced outages) of onshore wind turbines is normally less than 2% annually (Blanco, 2009).

The maximum wind power that can be extracted by a wind turbine is calculated using Equation 2.7 given as

$$P = \frac{1}{2}\rho A v^3 C_p \tag{2.7}$$

where P is the electrical power output from the wind turbine (W), ρ is the density of air (kg/m^3), A is the swept area of the turbine blades (m^2), V is the wind speed at which the turbine is rated (m/s) and C_p is the power coefficient of the turbine, which varies with the ratio of blade tip speed to wind speed.

The annual electrical energy generated by a wind turbine depends on the number of full load hours in the location where it is installed. It is calculated using Equation 2.8 given as

$$E_o = P_r \times t = P_r \times CF \times 8760 \tag{2.8}$$

where E_o is the energy generated (kWh/year), P_r is the rated power (kW), t is the number of full load hours in a year (h) and CF is the capacity factor (%).

The lifespan of wind turbines used in financial appraisals has been either 15 or 20 years with 20 years being increasingly used as the preferred bankable period. However, recent experience has shown that wind turbines could last 30 years or longer although they are most likely to be replaced with more advanced technology before this period. Thus, while financial appraisals of wind power plants are normally performed using 20 years, 25- or 30-year lifespans may be considered in sensitivity analysis. Typical turbine sizes range between 1.5 and 3.5 MW for onshore and 1.5 and 7.5 MW for offshore applications. There is a steady drop in average unit installed costs when moving from projects of 5 MW or less to projects in the 20–50 MW range. However, as project size increases beyond 50 MW, there is no evidence of continued economies of scale (E.ON, 2008).

Wind turbine operation is characterised by no fuel cost and relatively low annual O&M costs. The electricity generation cost for wind power plants is mainly influenced by the number of equivalent full load operating hours, capital and variable costs. In Europe, the generation costs of onshore wind farms range between 4.5 and 8.7 €c/kWh, while those for offshore wind farms range between 15 and 19 €c/kWh (IRENA, 2012b).

Annual O&M costs vary widely and have been reported to be between 1% and 10% of capital expenditure (Weaver, 2012) depending on factors such as technology characteristics, wind speed characteristics, site location and plant age. Variable costs fluctuate between 1 and 2 €c/kWh over the lifetime of a wind turbine (i.e. between 10% and 20% of the total costs) (Blanco, 2009). The capital cost of onshore and offshore wind is typically between 1100 and 1400 €/kW and between 1800 and 3400 €/kW, respectively (REN21, 2013). The greater cost of offshore wind is due to a wide variety of factors, including its relative technological immaturity, the significant extra civil works (foundations), difficult site access (requiring specialist boats, barges and equipment) and the additional costs of offshore cables and transformers.

The cost of generating electricity from onshore and offshore wind power plants is mainly influenced by capital costs, O&M costs and network connection costs, which can be very significant in offshore plants. Capital costs include wind turbines (production, blades, transformer, transportation to the site and installation), foundations and road construction. O&M of wind turbines (provision for repair and spare parts and maintenance of the electric installation) is the most important component of variable costs. Other variable costs include land and sub-station rental, insurance and taxes and management and administration (audits, management activities, forecasting services and remote-monitoring measures). For offshore plants, additional O&M costs include the cost of specialised staff and equipment (helicopters, vessels, cranes, etc.). Network connection costs relate to the installation of new high voltage transmission facilities (cables, sub-stations, connection and power evacuation systems).

Between 1996 and 2012, global wind turbine capacity increased approximately 45-fold, as shown in Figure 2.6. The technology learning rates for onshore wind energy between 2004 and 2010 were 9–19% (Wiser et al., 2011). Table 2.4 presents some key financial and technical characteristics of onshore and offshore wind power plants.

2.3.5 Ocean energy

Ocean energy is the energy captured from ocean waves, tides, salinity gradients and ocean temperature differences. Wave energy converters capture the energy of surface waves to generate electricity, tidal stream generators use kinetic energy of moving water to power turbines and tidal barrages are essentially dams that cross tidal estuaries and capture energy as tides flow in and out.

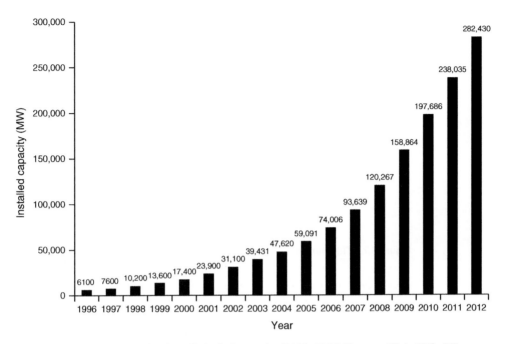

Figure 2.6   Global cumulative installed wind capacity (1996–2012) (*Source:* Global Wind Energy Council, 2013).

Table 2.4   Typical characteristics of onshore and offshore wind power plants.

Parameter	Onshore wind	Offshore wind
Turbine size (MW)	1.5–3.5	1.5–7.5
Capital cost (€/kW)	1100–1400	1800–3400
Service life (yr)	15–30	
Capacity factor (%)	19–40	35–45
LCOE (€c/kWh)	4–12	6–17
O&M costs (€c/kWh)	1–2	2–4
O&M costs (% of CAPEX)	10–20	10–30
Learning rate (%)	9–19	

Source: REN21 (2013), Blanco (2009), Wiser et al. (2011) and Weaver (2012).

There are four main types of wave-energy technologies classified on the basis of their operating principle. They include oscillating water columns; overtopping devices; floats, buoys or pitching devices and point absorbers. The vast majority of recently proposed wave-energy projects use offshore floats, buoys or pitching devices. Tidal dams or barrages have been the dominant type of tidal technology deployed to date.

The kinetic and potential energies present in ocean tides and waves represent a significant and useful energy resource if they can be exploited reliably and cost-effectively. Although many ocean energy conversion technologies

Figure 2.7   Tidal turbine.

have been developed, they are largely still at demonstration stage and have yet to achieve successful commercialisation. Low-head, propeller-type turbines are used to convert the energy of tides into electricity. High capital costs and limited suitable sites with adequate tidal ranges or flow velocities have hindered widespread development of this technology. However, there have been recent developments in both tidal power design and turbine technology including tidal stream generators (which use the kinetic energy of moving water), tidal barrage (which makes use of the potential energy resulting from the difference in height between high and low tides) and dynamic tidal power (which uses the interaction between kinetic and potential energies in tidal flows). Figure 2.7 shows a schematic of a tidal stream turbine. Turbine efficiency range between 30% and 50%, while the service life is estimated at between 15 and 25 years with capacity ratings of 100–200 kW.

Average worldwide tidal intervals are of 12 h and 24 min resulting in 706 tidal cycles per annum. Tidal turbines can harness the resulting tidal flows in a similar manner to wind turbines. The average annual electricity output from such a device can be expressed as

$$\overline{E}_{o} = 0.986 \times 10^{6} \times COP \times R^{2} \times A \tag{2.9}$$

where $\overline{E}_{o}$ is the average annual electricity generation (kWh), COP is the overall coefficient of performance (typically 0.2–0.35), R is the total range (m) and A is the area of the basin, that is, the water volume per unit height (km^{2}).

Wave power is similar to wind power in that it is intermittent and unpredictable, but differs from tidal and solar power, which are intermittent but are largely predictable. The average energy density per unit area of gravity waves on the water surface is expressed as

$$E = \frac{1}{8}\rho g H_{m0}^{2} \tag{2.10}$$

Table 2.5 Characteristics of tidal and wave energy power plants.

Parameter	Tidal energy	Wave energy
Plant size (MW)	1–260	1–50
Capital cost (€/kW)	3900–4400	
Capacity factor (%)	23–29	
LCOE (€c/kWh)	15–21	

Source: REN21 (2013); www.smartgrid.ieee.org.

where E is the mean wave density per unit horizontal area (J/m^2), ρ is the water density (kg/m^3), g is the acceleration due to gravity (m/s^2) and H_{m0} is the significant wave height (m).

The global installed capacity of ocean energy (mostly tidal) was 527 MW at the end of 2012, of which almost half (254 MW) was installed in 2011. Table 2.5 presents some key financial and technical characteristics of tidal and wave-energy power plants.

2.3.6   Photovoltaics

PV technology directly converts solar radiation to electrical energy using the photoelectric effect. Although PV energy output is intermittent, it is largely predictable due to the regular motion of the sun. However, cloud cover and ambient temperature are unpredictable and reduce system output. PV systems comprise several components including PV cells (generator), support structures, batteries or electricity grid (storage), power conditioning units and, sometimes, supplementary or backup generator (in stand-alone systems) to form a hybrid system. Power conditioning is essential as PV cells produce direct current (DC), while most appliances use alternating current (AC). Systems are rated in peak kilowatts (kW$_p$) for small systems and peak megawatts (MW$_p$) for larger systems, which represent the amount of electrical power that a system is expected to deliver under standard test conditions (solar intensity of 1000 W/m^2, PV cell temperature of 25°C and air mass ratio of 1.5).

PV modules are the principal electricity-generating component in PV systems and their efficiencies range between 10% and 30%, depending on the technology used. The overall efficiency of a PV system is a combination of the efficiency of the modules and balance-of-system (BOS) components such as inverters and cables. Ambient conditions also affect the performance as module efficiency decreases with higher temperatures. The efficiency of installed modules also decreases over time due to material degradation, and manufacturers usually guarantee a power output of 80% of their nominal power after 20–25 years. PV modules have a technology learning rate of $20 \pm 5\%$, which has resulted in substantial cost reductions and significant growth rate in the installed capacity of PV systems worldwide, as shown in Figure 2.9.

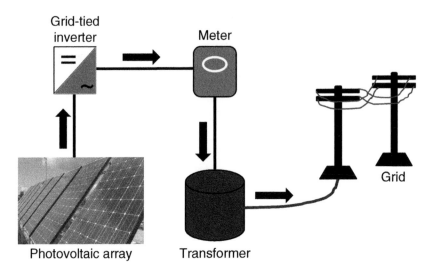

Figure 2.8   Grid-connected PV system.

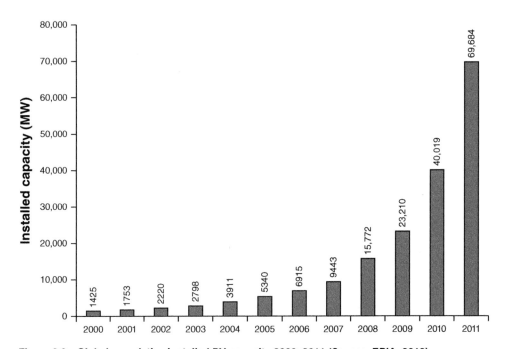

Figure 2.9   Global cumulative installed PV capacity 2000–2011 (*Source:* EPIA, 2012).

Cumulative installed capacities grew from 1425 to 69,684 MW between 2000 and 2011, respectively.

Grid-connected PV systems (see Figure 2.8) have a variety of configurations including large centralised power stations (often referred to as PV farms) and systems that are mounted on or integrated into commercial buildings and individual dwellings. When the PV system is used in buildings, the PV modules are mounted on either rooftops or façades, which can reduce the size and cost of mounting structures and land requirements. In building-integrated applications, PV materials are used to replace conventional building materials in the buildings envelope.

Centralised, grid-connected PV systems perform the functions of centralised power stations. The power supplied by such a system is not associated with a particular electricity customer, and the system is not located to perform specific functions in the electricity network other than the supply of bulk power. These systems are typically ground-mounted and function independently.

Utility scale, grid-connected PV plants use either stationary structures or sun-tracking systems in order to maximise energy production. Stationary structures are oriented at an optimal slope based on the latitude of the local site and are secured such that they provide enough resistance to mainly wind loads. Sun-tracking systems are either single- or dual-axis subject to economic evaluations as the tracking equipment increases project cost. The energy generated by PV systems is largely dependent on the intensity of solar radiation, which is intermittent in nature. It is, therefore, not suitable for supplying base loads. However, the attractiveness of PV plant has increased partly due to improved solar radiation forecasting methods.

Predicting the energy output from a PV system is a matter of combining the characteristics of the major components, that is, the PV array and the inverter with local solar insolation and temperature data. Assuming that the system efficiency remains almost constant throughout the day, the electrical energy generated can be estimated as

$$E_{o,t} = G_t \times A \times \eta_m \times \eta_{inv} \times T \tag{2.11}$$

where $E_{o,t}$ is the amount of alternating current electrical energy output in time interval t (kWh), G_t is the in-plane solar radiation (W/m^2), A is the PV array area (m^2), η_m is the PV module efficiency (%), η_{inv} is the inverter efficiency (%) and T is the time interval over which the energy output is calculated (hrs).

The capital costs of PV system components vary by size and application. For PV farms they include procurement and construction, engineering and developer costs. Procurement involves the purchase of PV modules, inverters and BOS components such as mounting structures and cables. It also covers expenses such as site preparation, system design, management, installer overhead, installation labour as well as the purchase of spares and contingencies. Engineering costs cover engineering design, construction management and commissioning. Developer costs include permitting, licencing, legal fees, grid interconnection, insurance, land acquisition and site preparation. For smaller building-mounted installations engineering and developer costs are normally low relative to the investment size.

Table 2.6 Selected economic and technical characteristics of PV systems.

Parameter	Value	
	Rooftop	Ground-mounted
Peak capacity	3–5 kW (residential) 100 kW (commercial) 500 kW (industrial)	2.5–250 MW
Installed cost (€/kW)	1700–3750	975–2100
LCOE (€c/kWh)	12–42	7–30
Service life (yr)	25–30	25–30
Annual O&M (% of CAPEX)	1–5	1–5
Efficiency (%)	10–30	10–30
Learning rate (%)	15–25	15–25
Capacity factor (%)	10–25	10–25

PV modules and BOS equipment usually accounts for about 80% of total system costs, with the PV modules covering the greatest share of the expenses. The remaining 20% is usually for installation, excluding a small amount for on-going maintenance costs (Stapleton and Neill, 2012). Investment costs depend on whether systems are rooftop or ground-mounted and the country where they are installed. They vary widely with average values ranging between 1700 and 3730 €/kW$_p$ for rooftop PV systems, while for ground-mounted PV systems, they vary between 975 and 2100 €/kW$_p$.

O&M costs normally include insurance, grid fees, labour costs for security, plant management, maintenance and material replacement costs. The most significant scheduled maintenance event is inverter replacement after approximately 10 years. O&M costs account for between 1% and 5% of investment costs. Table 2.6 details selected economic and technical characteristics of PV systems.

2.4 Heat generation

2.4.1 Boilers

A boiler is a closed fuel-burning container in which water or other fluid is heated to generate hot water, steam or vapour, superheat steam or any combination thereof for a variety of end uses. Boilers can be generally classified as pot or haycock, fire tube, water tube, flash, fire tube with water box and sectional. Figure 2.10 shows a schematic representation of a water-tube boiler. Boilers use a variety of fuels, which include fuel oil, natural gas, electricity and wood pellets or chips. Natural gas fired burners are used in most commercial building heating applications because natural gas is usually readily available, burns cleanly, and is typically less expensive than electricity or oil. Boilers have a long life, and the range of boilers available today have efficiencies greater than 90%.

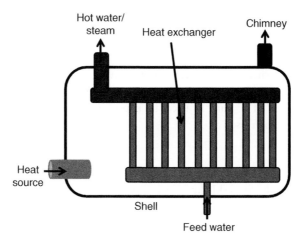

Figure 2.10   Water-tube boiler.

Fuel costs can be considerable with regular maintenance required. Delays in maintenance can result in costly repairs. Boiler operating costs can increase by about 10% for every year they go unattended.

Conventional boilers operate without condensing out water vapour from the flue gas. Condensing boilers on the other hand, operate at a lower return water temperature as compared to conventional boilers, which causes water vapour to condense out of the exhaust gases. This allows condensing boilers to extract additional heat from the phase change from water vapour to liquid, thereby increasing boiler efficiency. Conventional non-condensing boilers typically operate in the 75–86% (HHV) combustion efficiency range, while condensing boilers generally operate in the 88–95% (HHV) combustion efficiency range.

'Combi' or combination boilers exist that provide heat for central heating and domestic hot water on demand. They consist of a high efficiency domestic hot water heater and a central heating boiler combined within one compact unit. They reduce space requirements in buildings by eliminating the need for hot water cylinders (Table 2.7).

Table 2.7   Selected economic characteristics of some boilers.

Boiler	Capital cost (€/kW)	Operating cost excluding fuel (€/kW/annum)
Industrial gas boiler (large)	39–84	18–28
Commercial gas boiler (small)	~121	6–10
Commercial gas boiler (large)	~84	20–28
Industrial gas boiler (small)	39–84	21–28
Domestic gas boiler	163–194	~11
Non-domestic oil boiler (€/kWh)	23–69	2–5
Domestic oil boiler (€/kWh)	26–47	0.3–3

The heat output from a boiler is given as

$$Q_o = Q_i \times \eta \tag{2.12}$$

where Q_o is the heat output (kW), Q_i is the heat input from the fuel (kW) and η is the boiler efficiency (%).

If a fluid is used to export heat from the boiler, the heat output is expressed as

$$Q_o = \dot{m}C_p\Delta T \tag{2.13}$$

where $\dot{m}$ is the fluid mass flow rate (kg/s), C_p is the specific heat capacity of the fluid (kJ/(kg °C)) and ΔT is the temperature difference between the inlet and outlet fluid (°C)

2.4.2 Solar water heaters

Solar water heating systems (SWHSs) convert solar radiation into thermal energy, typically using water-based liquids as energy carriers. They are used in domestic, commercial and industrial applications for the production of hot water process heat. They are a mature technology and the most popular means of using solar energy. Thermosyphon or natural circulation and forced circulation or pumped systems are the two main types of SWHS. In a thermosyphon SWHS, the heat absorbed by the collector heats up water and causes its density to decrease. Denser cold water fed through the bottom of the collector then forces the warmer water to rise through natural circulation (thermosyphon effect) into an overhead storage tank. These SWHSs are suitable for sunny climates. Figure 2.11 shows a schematic of a thermosyphon SWHS. In a forced

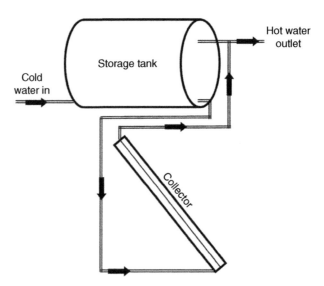

Figure 2.11 Thermosyphon SWHS.

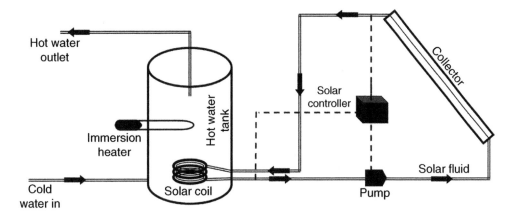

Figure 2.12   Forced circulation SWHS.

circulation SWHS, water is heated in the collectors and a pump is used to circulate a water-glycol mixture used as the heat transfer fluid. A solar controller triggers the pump when the solar fluid outlet temperature from the collector is above a set value compared to the water temperature at the bottom of the storage tank. A solar coil at the bottom of the hot water tank is used to heat water. The solar fluid should have some desirable properties such as low freezing and high boiling points. The ease of operation and low cost of solar energy collectors make them suitable for low temperature applications below 80°C. Often an auxiliary heating system (electric immersion heating) is used to raise the water temperature during periods when there is insufficient heat available from the solar collector. Systems with forced circulation are predominant in temperate climates, such as in central and northern Europe. Figure 2.12 shows a schematic of a forced circulation SWHS.

The basic components of SWHSs are collectors, storage tanks, connecting pipes, auxiliary heating systems and pumps (in forced circulation systems). Collectors are the main component and their optimal performance is crucial to the overall operation of SWHSs. They absorb both diffuse and direct solar radiation and therefore function even under overcast skies. They are distinguished by their motion, that is, stationary, single-axis tracking and dual-axis tracking as well as by their operating temperature. Stationary solar energy collectors are permanently fixed in position and do not track the sun. Three types of collectors fall into this category, namely, flat plate collectors (FPC), evacuated tube collectors (ETCs) and compound parabolic collectors (CPCs). FPCs and ETCs are widely used in solar thermal systems. Figures 2.13 and 2.14 show two FPCs and a heat pipe ETC, respectively. The efficiencies of SWHSs are typically around 30–35% for systems with FPCs and 45–50% for those with ETCs. The optimum tilt angle of the collector is equal to the latitude of the location, where it is installed with angle variations of 10–15° more or less depending on the application.

Figure 2.13   Flat plate collectors.

Figure 2.14   Heat pipe evacuated tube collector.

The basic method of evaluating the performance of a SWHS is to expose the collector to solar radiation and measure the incident solar radiation on the collector's surface and the solar fluid properties (i.e. flow rate, inlet and outlet temperature to the storage tank). The useful heat gain by the water in the tank is then calculated using Equation 2.14 given as

$$Q_u = \dot{m}_f C_p (T_{out} - T_{in}) \tag{2.14}$$

where Q_u is the useful heat gain (kW), C_p is the heat capacity of the solar fluid (kJ/(kg °C)), $\dot{m}_f$ is the solar fluid mass flow rate (kg/s), T_{out} is the solar fluid temperature at the tank outlet (°C) and T_{in} is the solar fluid temperature at the inlet to the tank (°C).

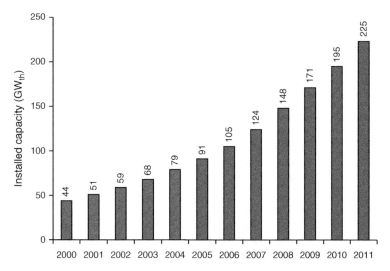

Figure 2.15 Global installed capacity of glazed SWHSs. (*Source*: REN21, 2013).

Solar water heating systems are used to generate hot water and displace the heat generated by conventional heating system that may be based on fossil fuels or electricity. The solar fraction (SF) is the ratio of solar heat yield to the total water heating energy requirement in the building and is expressed as

$$SF = \frac{Q_u}{Q_u + Q_{aux}} \qquad (2.15)$$

where SF is the solar fraction (dimensionless), Q_u is the useful heat gain from solar energy (kWh) and Q_{aux} is the auxiliary heat (kWh).

Their economic viability is, therefore, evaluated on the basis of their additional capital cost and the annual cost savings resulting from the heat produced. This is highly dependent on the cost of the displaced fuel.

The installed cost of SWHS varies from 825 to 1650 €/kW$_{th}$ for single-family homes and from 710 to 1550 €/kW$_{th}$ for multi-family homes. For district heating schemes, the installed cost varies from 345 to 585 €/kW$_{th}$ for systems without storage and from 350 to 800 €/kW$_{th}$ for systems with seasonal storage. The relationship between collector area and capacity is $1\ m^2 \approx 0.7\ kW_{th}$ (kilowatt-thermal). Annual O&M costs are approximately 1% of investment costs while the system useful life is typically 20 years. Table 2.8 details the characteristics of thermosyphon and forced circulation SWHSs.

SWHSs are increasingly contributing to domestic hot water provision and space heating in many countries. Global annual installations have been on a steady rise from 44 GW$_{th}$ in 2000 to 233 GW$_{th}$ in 2011. The global market grew by 15% between 2010 and 2011. Figure 2.15 shows the evolution of global installed capacity of glazed SWHSs between 2000 and 2011.

Table 2.8 Typical characteristics of thermosyphon and forced circulation SWHSs.

Parameter	Value	
	Domestic hot water	**Domestic heat and hot water**
Annual O&M (% of investment cost)	1	1
Service life (yr)	20	20
Learning rate (%)	10	10
System efficiency (%)		
FPC	30–35	30–35
ETC	45–50	45–50
Plant size (kW_{th})		
Single family	2.1–4.2	7–10
Multi-family	35	70–130
District heating		70–3500
District heating with seasonal storage		>3500
Installed cost (€/kW_{th})		
Single family	825–1650	825–1650
Multi-family	710–1550	710–1550
District heating		345–585
District heating with seasonal storage		350–800
Energy cost (€c/kWh)		
Domestic hot water	1.1–21	3.8–38
District heat		>3

Source: REN21 (2013).

2.5 Combined heat and power

Combined heat and power (CHP) is the simultaneous generation of electrical and thermal energy from the same source: fuel is burned to drive a generator and produce electricity and the heat produced in this process is recovered and used for heating. It results in the improved use of the available fuel with conversion efficiencies over 70% and up to 90%. CHP plants can be powered by a wide variety of fuels, including natural gas, biogas or diesel (gas oil). They operate best on sites that have a year-round heat demand and are typically economically viable operations when they run for more than 5000 h annually. A variety of size classifications exist, but they can be characterised by thresholds of electrical output as micro combined heat and power (micro-CHP) (up to 5 kW(e)), small-scale CHP (5 kW(e) to 2 MW(e)) and large-scale CHP (above 2 MW(e)). CHP technologies fall into three groups: internal combustion engines, external combustion engines (e.g. Stirling cycle) and fuel cells.

CHP plants are normally embedded in a site displacing the need for imported electricity and displacing boiler fuel. They are generally operated as electricity- or heat-led modes. In electricity-led mode, the CHP plant is operated to meet the site's power demand. In heat-led mode, it is operated to meet the heat demand. Top up of electricity and fuel is needed during periods of high demand as well as forced and planned outages. Excess electricity generated is spilled or exported to the electricity distribution network. Top up heat is required when heat generation is less than demand; heat is either stored or dumped when there is insufficient demand. Although the overall efficiency of CHP systems is similar to that of boilers, the electricity produced by the former has a much higher value than heat. It is the value of this electricity which must largely cover the investment cost of the CHP unit to provide a net saving.

Analysing the economic performance of a CHP plant involves comparing its performance against that of the alternative heating system retrofitted in the same facility and the import of grid electricity. In the case of a building connected to the electricity grid, it is normally assumed that the grid will meet the top-up electrical energy demand or supply the whole electrical demand if the CHP is not running, while for the alternative heating system solution, all electricity used in the building is imported from the grid.

There are three different system efficiencies that are important in measuring the energy performance of CHP systems: the system thermal efficiency, the system electrical efficiency and the system overall efficiency. These are normally specified by the manufacturer, although these efficiencies are often higher than those achieved in practice. The CHP unit's overall efficiency is the ratio of total electrical and thermal energy produced to the total fuel energy (HHV) supplied to the system. This value is found by adding the electrical efficiency and the thermal efficiency and is expressed as

$$\eta_{overall} = \frac{E_{el} + Q_h}{Q_f} = \eta_{el} + \eta_{th} \qquad (2.16)$$

where $\eta_{overall}$ is the overall efficiency (dimensionless), Q_f is the quantity of fuel used (kWh), Q_h is the quantity of heat generated (kWh), E_{el} is the electrical energy generated (kWh), η_{el} is the electrical efficiency and η_{th} is the thermal efficiency.

During operation, for a given quantity of fuel used, a CHP unit generates both heat and electrical energy while incurring some energy losses. This is expressed mathematically in Equation 2.17 as

$$Q_f = Q_h + E_{el} + E_{loss} \qquad (2.17)$$

where E_{loss} is the energy loss (kWh).

For a given quantity of fuel used and a specified thermal efficiency, the quantity of heat generated is expressed as

$$Q_h = Q_f \times \eta_{th} \qquad (2.18)$$

For a given quantity of fuel used and a specified electrical efficiency, the quantity of electricity generated is expressed as

$$E_{el} = Q_f \times \eta_{el} \qquad (2.19)$$

2.5.1 Micro-CHP

Micro-CHP refers to a group of technologies that generate both usable heat and electricity on a small scale in and around buildings. The heat generated is used for space and hot water heating, while the electricity generated is used on-site or exported to the national electricity grid. The electricity generated displaces that generated at central power stations resulting in a reduction in transmission and distribution losses as well as carbon emissions. Micro-CHP units are especially suited to situations where there is simultaneous heat and power demand as is prevalent in houses in northern European countries. Typical operating efficiencies for micro-CHP units are between 80% and 90% as compared with less than 40% of primary energy input for conventional power-generating systems that supply electricity to the grid. Figure 2.16 shows the energy flows in a domestic dwelling with a micro-CHP unit.

The system parameters required to evaluate the economic performance of a micro-CHP include grid electricity import and export tariffs, discount rates, economic lifespan, O&M costs, electrical and thermal efficiencies, gas (or other fuel) costs and installation costs. The heating system parameters required include capital and installation costs, O&M costs, system efficiency, economic lifespan and fuel costs. Table 2.9 details the characteristics of different micro-CHP technologies.

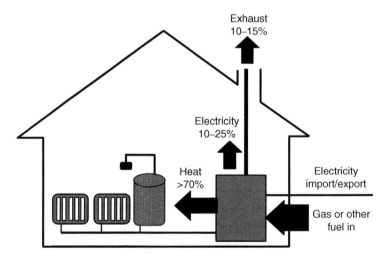

Figure 2.16 **Energy flows in a domestic dwelling with a micro-CHP unit.**

Table 2.9   Typical characteristics of micro-CHP technologies.

Parameter	Stirling engine	Internal combustion engine	Fuel cell (PEM)	Fuel cell (solid oxide)
Electrical output range (kW(e))	1–100	1–15	1–250	1– > 1,000
Electrical efficiency (%)	12–27	27–32	30–45	40–50
Power to heat ratio	1:3 to 1:8	1:1 to 1:2.2	1:2 to 1:1	1:1 to 2:1
CHP efficiency (%)[a]	Up to 90	Up to 90	Up to 90	Up to 90
Capital cost (€/kW(e))	800–2,500	1,500–2,000	2,500–3,000	4,000
Installed cost (€/kW(e))	1,000–3,000	1,800–2,300	2,750–3,500	4,500
Lifetime (h)	30,000	80,000	80,000	40,000

[a] Based on gross calorific value.
Source: Delta Energy and Environment (2006).

2.5.2   CHP engines

Reciprocating engines are the oldest CHP technology developed over a century ago and have been widely used for electricity generation. Both Otto (spark ignition) and Diesel cycle (compression ignition) engines have been used for applications ranging from fractional horsepower units for small handheld tools to several MW base-load electric power plants. CHP applications have focused primarily on natural-gas-fired spark ignited (SI) units in the 60 kW to 4 MW size range (Table 2.10). Most of the larger sizes of natural-gas-powered CHP units use lean-burn technology that results in increased efficiency and lower emissions (LeMar, 2002).

Table 2.10   Economic characteristics of CHP reciprocating engines.

Parameter	Large scale	Small scale
Capacity (kW(e))	100–3000	1–100
Service life (yr)	15–20	15–25
Electrical efficiency (%)	30–40	20–40
Overall efficiency (%)	75–85	80–85
Installed cost (€/kW(e))	770–1250	1150–9250
Fixed O&M (€/(kW(e) yr))	1.2–7.7	Varies
Variable O&M (€/kWh)	0.006–0.013	0.008–0.013

Source: OECD/IEA (2010).

2.5.3   CHP turbines

For several decades, combustion turbines ranging from 1 MW to several hundred MW have been used for power generation when configured as a combined cycle power plant. Industrial turbines are generally in the 1–15 MW

Table 2.11 Selected economic and technical characteristics of CHP technologies.

Parameter	Steam turbine	Gas turbine	Micro-turbine	Reciprocating engine	Stirling engine	Fuel cell
Capacity (kW)	100–250,000	250–250,000	30–250,000	0.5–5,000	2–1,250	0.5–2,000
Electrical efficiency, HHV (%)	15–38	22–36	25–40	26–40	15–30	30–63
Overall efficiency, HHV (%)	~80	70–80	70–85	70–92	75–95	80–90
Power-to-heat ratio	0.1–0.3	0.5–2	0.4–0.7	0.5–1		1–2
Installed cost (€/kW(e))	350–850	750–1,000	1,850–2,300	650–1,700	850–2,000	3,850–5,000
O&M cost (€/MW(e))	<3.85	3.08–8.46	9.23–19.23	6.92–16.92	6.92–10.00	7.54–11.31
Capacity factor (%)	~100	90–98	90–98	92–97		>95

Source: US EPA (2008).

range with larger units used by utilities, while smaller units are referred to as microturbines. Combustion turbines are characterised by relatively low installed cost, low emissions and infrequent maintenance. They are typically used for CHP when a continuous supply of steam or hot water and power is desired. Table 2.11 details selected economic and technical characteristics of CHP technologies.

2.5.4   Combined heat, power and cooling

Combined heat, power and cooling (CHPC) works by the same principles as conventional CHP except the system is extended to drive absorption chillers for cooling applications. Absorption chillers use thermodynamic heat pump principles to produce chilled water from a heat source, much in the same way that a refrigerator works. They can use waste heat from a CHP (when end-use heating demand is low) to improve overall system efficiencies and economies. Cooling can be used in commercial offices and industrial applications such as food refrigeration and data centre cooling. The waste heat from the CHP system is used to boil a solution of refrigerant/absorbent which is then captured and used to chill water after a series of condensation evaporation and absorbption steps are performed. This process is essentially a thermal compressor, which replaces the electrical compressor in a conventional electric chiller. In doing so, the electrical requirements are significantly reduced, requiring electricity only to drive the pumps that circulate the solution. Single-effect units offer coefficient of performances (COPs) of about 0.7, whereas double-effect units attain levels of about 1.2. Double-effect units, however, require a higher

Table 2.12   Cost and performance of single-effect indirect-fired absorption chillers and engine-driven chillers.

Tons	Chiller	Cost (€/ton)	Electric use (kW/ton)	Thermal input (kWh/kg)	Maintenance cost (€/ton annual)
10–100	SEIFAC	500–950	0.02–0.04	4.98–5.57	23–62
	EDC	600–850	0.05–0.07	2.64–3.52	35–77
100–500	SEIFAC	300–550	0.02–0.04	4.98–5.27	15–38
	EDC	500–750	0.01–0.05	2.34–3.22	27–58
500–2,000	SEIFAC	200–400	0.02–0.05	4.98–5.27	8–23
	EDC	350–600	0.003–0.01	2.05–2.34	19–46

Abbreviations: SEIFAC, single-effect indirect-fired absorption chiller; EDC, engine-driven chiller.
Source: LeMar (2002).

temperature source that cannot be provided by some CHP systems, particularly smaller reciprocating engines, turbines and fuel cells. The COP is defined as the ratio of cooling output to fuel input.

Engine-driven chillers (EDCs) are conventional chillers driven by an engine instead of an electric motor. They employ the same technology that electric chillers use, but use a gas-fired reciprocating engine to drive the compressor. As a result, EDCs can be economically used to provide cooling where gas tariffs are relatively low and electricity tariffs are high. Another benefit offered by EDCs is the better variable speed performance, which results in improved part load efficiencies. EDCs operate in a CHP system when the waste heat produced by the engine is recovered, and used for space heating and/or domestic hot water loads (Table 2.12).

2.6   Energy storage

Cost-effective energy storage has become a pivotal issue in the move to decarbonising energy systems. Because so many renewable energy systems are intermittent, they result in a mismatch between supply and demand, thus requiring conventional technologies to switch on and off and vary their output more frequently than before. Many conventional technologies, however, are not designed to meet these operating conditions. Storage can alleviate this problem by providing additional demand (during charging) and supply (discharging) when needed.

The economic rationale for energy storage systems is based on

- the storage of surplus energy, which would otherwise be wasted for later productive use;
- arbitrage, where storage occurs at times of low energy prices and energy is discharged during periods of high prices; and
- the provision of electricity system services, which maintain system stability such as load levelling to correct imbalances in supply and demand.

Storage technologies store energy in many forms such as electrical (batteries), mechanical (flywheels), chemical (hydrogen), potential (pumped hydro) and thermal (phase change materials). Their economics depends on capital costs and price arbitrage. In liberalised electricity markets, bulk energy storage projects may be remunerated through capacity payments, electricity trading and ancillary services payments.

The round-trip efficiency of an energy storage system is related to the energy input and output as

$$\eta = \frac{E_o}{E_i} \qquad (2.20)$$

where η is the round-trip efficiency, E_o is the electricity output (kWh) and E_i is the electricity input (kWh).

Only technologies suitable for large scale electrical and thermal energy storage will be discussed in this section because these are most relevant to the objectives of the book.

2.6.1 Electrical

Energy storage technologies which can be used on electrical networks include batteries, flow batteries, fuel cells, flywheels, superconducting magnetic energy storage (SMES), super capacitors, compressed air energy storage (CAES) and pumped hydro. Pumped hydro storage is a mature technology and the oldest and largest of these technologies. CAES systems are also suitable for large scale energy storage.

For many applications, a few hours of storage is sufficient to reduce peak demand and capture significant value; however, beyond several hours, this marginal value decreases significantly. The economics of electrical energy storage systems depends on capital cost, time of charging and discharging, round-trip efficiency, energy tariffs, and the ability to provide system services. Investment costs vary widely depending on the plant design and storage technology used.

2.6.2 Pumped hydroelectric storage

Pumped hydroelectric storage (PHS) is the most used large-scale (>100 MW) energy storage system. Its operating principle is based on managing the gravitational potential energy of water by pumping it from a lower reservoir to an upper reservoir during periods of low power demand, as shown in Figure 2.17. When power demand is high, water flows from the upper reservoir through the intake shaft and power tunnel to the lower reservoir, activating the turbines in the power house to generate electricity. The energy stored is proportional to the water volume in the upper reservoir and the height of the waterfall (Rabiee et al., 2013). In general, PHS plants have a round-trip efficiency of 55%–85%. The major drawbacks are limitations in suitable

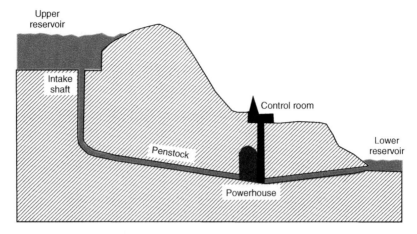

Figure 2.17 Pumped storage hydropower plant.

sites for widespread deployment, long lead times, high capital costs and environmental issues.

The electrical energy generated by a PHS plant in a year depends on the potential energy of the water contained in the upper reservoir less system losses in the penstock, turbine and generator; the capacity factor; and the number of hours in a year. The annual electrical energy generated by a PHS plant is calculated using Equation 2.6 for hydropower given in Section 2.3.3.

PHS plants are characterised by long asset lives (typically 50–100 years), high capital costs and low O&M costs. Project costs for PHS are very site specific with some quoted costs varying from 600 to 3000 €/kW (Gatzen, 2008). Furthermore, capital costs depend not only on the installed power but also on the energy storage capacity at any given site (Deane et al., 2010). Arbitrage and system service provision are normally major sources of revenue. Table 2.13 details some general characteristics of PHS plants.

Data on the annual installed capacity of PHS plants is limited. The IHA (2010) reported a global total installed capacity of 136 GW. The IHA (2013) also reported that new PHS plants with a total of 2–3 GW were commissioned

Table 2.13 General characteristics of PHS plants.

Parameter	Values
Power rating (MW)	5–2,000
Discharge duration (h)	4–100
Efficiency (%)	55–85
Service life (yr)	50–100
Durability (cycles)	12,000–30,000+
Installed cost (€/kW)	450–2,500
Capacity factor (%)	95+
O&M cost (% of investment cost)	1–2

Source: Deane et al. (2010), Gatzen (2008), IHA (2013) and NREL (2012).

in 2012. Growth in the use of PHS plants has largely been as a result of their ability to provide ancillary services to grid operators as the market for variable renewable generation increases.

2.6.3 Compressed air energy storage

Compressed air energy storage (CAES) systems use electrical compressors to store air in underground caverns (typically at 4.0–8.0 MPa), which is later extracted and combusted with natural gas to drive gas turbines and generate electricity. Storage typically occurs during periods of low electricity demand, when there is excess electricity production (e.g. by wind farms during very windy periods) or when electricity prices are low. During electricity generation, the compressed air is drawn from the storage cavern, heated and then expanded in a set of high and low pressure turbines which convert most of the energy of the compressed air into rotational kinetic energy. In addition, the air is mixed with natural gas and combusted. Waste heat is used to preheat the incoming air supply. A CAES system schematic is shown in Figure 2.18.

CAES systems are suitable for bulk energy storage and are also used to provide ancillary services such as load levelling, spinning reserve, peak shaving, valley filling, contingency service, area control and black start. High capital cost, slow response and limitations in the suitability of sites are major

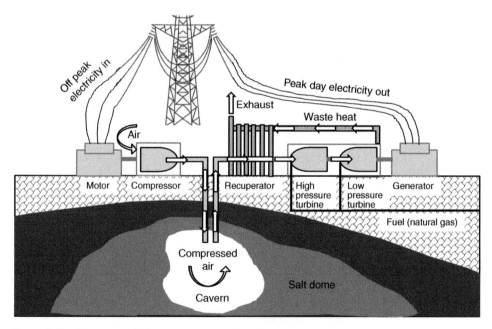

Figure 2.18 Compressed air energy storage in salt cavern.

disadvantages of the technology. The energy output from CAES plants is calculated as

$$E_o = E_i \times \eta_{ER} \quad \text{if } t_{comp,on} \leq t \leq t_{comp,off} \qquad (2.21)$$

where E_o is the CAES electrical energy output (MWh), E_i is the compressor electrical energy input (MWh), η_{ER} is the energy ratio or round-trip efficiency (plant electrical output/electrical input) and $t_{comp,on}$ and $t_{comp,off}$ denote time periods during which the compressor does and does not operate, respectively.

The performance of a CAES plant is dependent on the type of compressor and turbine used as well as the storage capacity of the cavern. CAES plants have very high ramping rates and the key to maintaining their efficiency properties is to continuously run them on part load operation. They can, therefore, provide power when needed to mitigate the effect of intermittent sources such as wind turbines and PV plants. Cavern maintenance is dependent on operational pressures.

The main CAES investment costs include land, cavern construction, compression and generation plant, commissioning, financing and interconnection and are highly dependent on the underground storage conditions and typically range between 300 and 600 €/kW. O&M costs include electricity (for compression), natural gas (for generation), maintenance, insurances, staff and other overheads. Underground CAES storage systems are most cost-effective with storage capacities up to 10 GWh, whereas overground units are typically smaller and more expensive, with capacities in the order of 60 MWh.

Revenue from CAES plants is derived from electricity sold to grid operators at the system marginal price (in a pool market) as well as capacity payments for the provision of ancillary services. Table 2.14 details some general characteristics of CAES plants.

Table 2.14   General characteristics of CAES.

Parameter	Value
Power rating (MW)	25–2,500
Discharge duration (h)	2–24
Efficiency (%)	40–70
Service life (yr)	15–40
Durability (cycles)	30,000+
Installed cost (€/kW)	450–1,000
Capacity factor (%)	65–96
O&M cost (% of investment cost)	0.6–2
Energy ratio	1.5
Heat rate (MJ/MWh)	4.2

Source: OECD/IEA (2010) and StoRE (2012).

2.6.4 Thermal energy storage

Thermal energy storage (TES) consists of storing energy as heat either at low or high temperatures. When needed the stored heat is either used directly or converted to electricity using heat engines. Thermal energy can be stored using either the heat capacity or the latent heat (resulting in phase change) of the storage material. Some common materials used for thermal energy storage include water, molten salts and lithium fluoride. Some storage systems use chilled water or ice, charging during the night and discharging during the day when energy is required for chilling.

The economic justification for TES systems assumes that the annual income needed to cover capital and operating costs be less than that required for primary generating equipment supplying the same service loads and periods.

TES systems aim to reduce primary energy use and thus both conserve fossil fuels and reduce GHG emissions. Energy savings can be achieved in several ways:

- system fuel use can be reduced by storing waste or surplus thermal energy available at certain times for use at other times;
- electrical energy costs can be reduced by storing electrically generated thermal energy during cheap off-peak-periods to meet the thermal loads that occur during high demand periods; and
- the purchase of additional heating, cooling or air-conditioning capacity can be avoided because stored energy can be withdrawn to meet maximum thermal loads that exceed existing capacity.

Sensible heat storage

Sensible heat storage (SHS) consists of storing thermal energy by raising the temperature of a solid or liquid. An SHS system uses the heat capacity and temperature change of the material during the charging and discharging process. SHS is mainly used for low grade heat such as solar energy or waste heat from power generation plants and industrial thermal processes for short- and long-term storage purposes in buildings. Water, oil, molten salts, molten metals, bricks, sand and soil are commonly used for short-term SHS while large aquifers, rock beds, solar ponds and large tanks are used for long-term (e.g. annual) storage.

The quantity of heat stored depends on the specific heat of the medium used, its change in temperature and the quantity of material used. The quantity of heat stored is calculated using Equation 2.22 given as

$$Q = MC_p \int_{T_i}^{T_f} \Delta T = MC_{p,avg}(T_f - T_i) \qquad (2.22)$$

where C_p is the specific heat of the storage material (kJ/(kg K)), $C_{p,avg}$ is the average specific heat between T_f and T_i (kJ/(kg K)), M is the mass of the storage material (kg); Q is the heat stored (kW), ΔT is the temperature change (K), T_f is the final temperature (K) and T_i is the initial temperature (K).

Latent heat storage

Latent heat storage (LHS) is based on the absorption or release of heat when a storage material undergoes a phase change from one physical state to another. Phase change occurs in: solid–liquid, solid–gas, liquid–gas and vice versa. LHS systems consist of three essential components: a phase change material (PCM) with its melting point in the desired temperature range, a heat exchange surface and a container compatible with the PCM. PCMs are usually classified as organic (paraffin, fatty acids, alkanes) or inorganic (salts).

LHS using PCMs is one of the most efficient methods to store thermal energy. The use of PCM provides higher heat storage capacity and more isothermal (constant temperature) behaviour during charging and discharging compared to sensible heat storage. High energy storage density and high power capacities for charging and discharging are desirable properties of any storage system. PCM systems can be used in domestic hot water tanks, space heating and cooling of buildings, peak load shifting, solar energy applications and seasonal storage, low temperature cooling (chilled water storage and ice storage systems) as well as other high temperature applications.

The energy storage capacity of an LHS system with a PCM material can be calculated using Equation 2.23 or 2.24 given as

$$Q = MC_p \int_{T_i}^{T_m} dT + Ma_m \Delta h_m + MC_p \int_{T_m}^{T_f} dT \qquad (2.23)$$

$$Q = M(C_{p,avg1}(T_m - T_i) + a_m \Delta h_m + C_{p,avg2}(T_f - T_m)) \qquad (2.24)$$

where Q is the heat stored (kW), M is the mass of the heat storage medium (kg), $C_{p,avg1}$ is the average heat capacity of the medium between T_i and T_m (kJ/(kg K)), a_m is the fraction melted, Δh_m is the heat of fusion per unit mass (kJ/kg), $C_{p,avg2}$ is the average heat capacity of the medium between T_m and T_f (kJ/(kg K)), T_m is the melting temperature (K), T_i is the initial temperature (K) and T_f is the final temperature (K) (Table 2.15).

2.7   Energy efficiency

Energy efficiency involves reducing the amount of energy used to deliver the same products and services. It leads to a reduction in energy costs resulting in

Table 2.15 Selected economic and technical characteristics of thermal storage technologies.

Technology	Storage capacity (kWh/ton)	Efficiency (%)	Initial investment cost (€/kW)	Heat of fusion (kJ/kg)	Latent heat (MJ/m^3)
Underground thermal energy storage	10–50	50–90	2,600–3,500		
Pit storage		50–90	75–230		
Molten salts		40–93	300–550		
Ice storage		75–90	4,600–11,600		
Thermochemical	≤250	80–99	750–2,300		
Salt hydrates/eutectics				150–270	240–580
Paraffins/waxes				150–250	150–190
Fatty acids				150–200	130–190

financial cost savings where these savings should sufficiently compensate for the additional cost of implementation. It also results in carbon dioxide emissions reductions. There is a very wide range of technologies and interventions that can be implemented to reduce energy use, including

- upgrading energy conversion plant, such as replacing an old boiler with a more modern condensing alternative;
- replacing energy end-use appliances, such as lighting systems;
- improving the performance of building fabric (insulation and glazing upgrades);
- changing individuals' behaviours;
- replacing energy controls; and
- improving management performance.

Here, however, we will only consider insulation and high efficiency lighting since these are used in case studies later in the text.

2.7.1 Thermal insulation

Thermal insulation is the reduction of the transfer of thermal energy between surfaces at different temperatures either in thermal contact or in a range of radiative influence. Building insulation materials are thermal insulation materials used in the construction or retrofit of buildings. There are different types of insulation available. Some products are suited for use in cavity walls, others for timber framed construction; some are load-bearing and suitable for use under concrete floors, while others are soft and better suited to fitting in the attic space. Insulation layers can either be applied on the outside or inside of buildings, the latter shown in Figure 2.19.

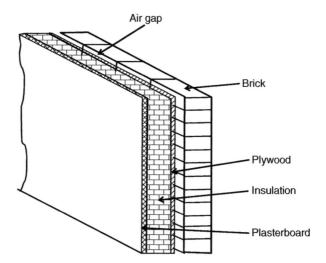

Air gap

Brick

Plywood

Insulation

Plasterboard

Figure 2.19   Building envelope thermal insulation.

Insulators are used to reduce thermal energy losses. Thermal conductivity is the most important property of insulating materials with a lower value providing a higher level of insulation for a given thickness. The U-value of an insulating material is the rate at which thermal energy is conducted through a unit area, per Kelvin temperature difference between its two sides. It is expressed as

$$U = \frac{Q_L}{A \times \Delta T} \tag{2.25}$$

where U represents the unit heat loss rate (W/(m^2 K)), Q_L is the rate of heat loss (W), A is the surface area (m^2) and ΔT is the temperature difference (K).

A good insulating material has a low U-value. Adding insulation to different areas of a building results in a reduction in the U-value and consequently heat loss. Table 2.16 details some typical U-values for different areas of a house with and without insulation. R-value measures thermal resistance and is the reciprocal of U-value.

The energy savings arising from upgrading the insulation of a surface is given as

$$E_s = \Delta U \times Q \tag{2.26}$$

where E_s is the energy saving (W), ΔU is the improvement in U-value (W/(m^2 K)); Q_L is the rate of heat loss (W); and A is the surface area (m^2).

The economics of an insulation upgrade is assessed by evaluating the energy savings resulting from an improvement in the U-value of the surface area over which the insulation is installed. The economic benefit of insulation varies for different applications and the method of financial appraisal used. Payback period (PP) (described in Section 4.3.1) is the time in years required to recover

Table 2.16   Typical *U*-values for different parts of a house.

	U-value (W/(m² K))
Roof	
Without insulation	2.3
With insulation	0.4
Cavity wall	
Without insulation	1.6
With insulation	0.6
Floor	
Without insulation	0.9
With insulation	0.6
Solid wall (block and plaster)	2.3
Cavity wall (block/cavity/block)	1.6
Cavity wall (block/polystyrene board/block)	0.58
Cavity wall (timber frame with above)	0.46
Window (single glazed, metal frame)	5.7
Window (double glazed −20 mm airspace, metal frame)	2.8

Table 2.17   Typical payback periods for energy efficiency measures in a traditional solid-wall northern-European house.

Measure	Payback time (yr)
Hot water lagging jacket	0.5
Draught proofing windows and doors	1
Roof insulation	1–2
Wall insulation	3
Adding an outside porch	30
Double glazing	40

a project's investment is also used to evaluate the economics of insulation projects. Table 2.17 details some common domestic insulation measures and their representative PPs.

2.7.2   High-efficiency lighting

Lighting is important in creating a pleasant environment in buildings and other open spaces. Good lighting offers several benefits such as improved productivity in work places and reduced accidents. It delivers visual comfort, good visibility, good colour reproduction where required, uniformity of light and minimises glare. Different types of lighting are now available that offer both energy and maintenance savings due to their energy efficiencies and longer life expectancies, respectively. Lighting types include incandescent, fluorescent (tube and compact), tungsten halogen, sodium high pressure, light emitting diode (LED), metal halide and mercury vapour. Table 2.18 details selected technical and economic characteristics of light bulbs.

Table 2.18 Selected technical and economic characteristics of light bulbs.

Type	Wattage (W)	Capital cost	Relative operating costs	Efficacy (lumens/W)	Average life (h)	Colour temperature (K)	Colour rendering (Ra)
Incandescent	15–1,500	Low	High	10–17	750–2,500	2,600	100
Tungsten halogen	20–2,000	Low	High	22	2,000	2,500	100
Fluorescent tubes (T5/T8)	8–36	Low/medium	Low	92–104	7,000–24,000	2,700–6,500	58–95
Compact fluorescent	9–52	Low/medium	Low	50–70	10,000	2,700–4,000	85
Metal halide	35–3,500	High	Very low	60–115	5,000–20,000	3,000–6,000	65–85
Mercury vapour	40–1,000	Low	High	25–60	16,000–24,000	3,300–4,200	36–58
High pressure sodium	35–3,500	High	Medium	50–140	16,000–24,000	2,000–2,200	25–65
Light-emitting diodes	3–4.2	High	Very low	10–100	50,000	3,000–6,000	70/80

Incandescent bulbs are extremely wasteful as about 90% of the electricity they use produces heat rather than light. Tungsten lights use less than 10–20% of the energy of, and last about twice as long as incandescent bulbs. They, however, generate heat and must not be used near flammable materials. Fluorescent lamps are efficient using about 20% of the power of incandescent bulbs. They are long lasting and generate little heat. Metal halides provide bright white point light. They are more efficient than mercury vapour and brighter than sodium lights. Mercury vapour lamps have a longer lifespan to metal halides. Sodium high pressure lamps produce a warm white light and have a longer lifespans than metal halide lamps. LEDs are semiconductor devices that are very energy efficient and produce very little heat. LEDs have higher initial cost but lower energy and maintenance costs compared to fluorescent technologies. This may however change in the future due to the rising costs of rare metals used to manufacture lamps.

Lamp life is the average elapsed time until replacement is necessary. It has a significant impact on economic performance since maintenance (labour) costs are an important consideration in commercial and industrial lamp replacement projects. LEDs typically have the longest lamp lives. Lumen is a measure of the quantity of light emitted by a lamp. Illuminance describes the light level on a particular surface and is measured in 'Lux'. It depends on the luminary placement, its light output intensity, and its light placement. Efficacy is the ratio of light emitted by a lamp to the power consumed by it, that is, lumens per watt.

The amount of energy consumed by lighting can be reduced by installing energy-efficient lighting and controls, optimising existing controls, making the most of natural lighting and observing good housekeeping practices and reducing lighting to the minimum required standards. The financial performance of energy efficient lighting systems depends on the cost of installing the new system as well as energy and maintenance cost savings. The payback period described in Section 4.3.1 or the savings-to-investment ratio (SIR) described in Section 4.3.3 can be used to assess the financial performance of an energy-efficient lighting investments. The annual energy cost saving arising from upgrading light bulbs from a higher power rating to a more efficient bulb with lower power rating is calculated using the following equation.

$$C_s = \left(\frac{(N \times \Delta P \times H \times T)}{1000} \right) \tag{2.27}$$

where C_s is the annual energy cost savings (€), N is the number of lamps replaced, ΔP is the difference in power rating (W), H is the usage per year (h/year), T is the electricity tariff (€/kWh) and C_R is the cost of replacing the lighting (€).

References

Blanco, M.I. (2009) The Economics of Wind Energy. *Renewable and Sustainable Energy Reviews*, **13**(6–7), 1372–82.

BP (2013) Statistical review of world energy 2013. Available at www.bp.com. Accessed 22 Oct 2014.

Delta Energy & Environment (2006) *New technologies for CHP applications*. Sustainable Energy Ireland, Dublin.

Deane, J., ÓGallachóir, B. and McKeogh, E. (2010) Techno-economic review of existing and new pumped hydro energy storage plant. *Renewable and Sustainable Energy Reviews*, **14**(4), 1293–1302

E.ON (2008) Scroby sands offshore wind farm, 3rd Annual Report: January 2007– December 2007.

EPIA (2012) Global market outlook for photovoltaics until 2016. Available at www.epia.org. Accessed 22 Oct 2014.

Gatzen, C. (2008) *The Economics of Power Storage*. Oldenburg Industriever, Munich.

Global Wind Energy Council (2013) Global wind statistics 2012. Available at www.gwec.net. Accessed 22 Oct 2014.

IHA (2010) Activity report: 2010. International Hydropower Association. Available at www.hydropower.org. Accessed 22 Oct 2014.

IHA (2012) Activity report: 2012. International Hydropower Association. Available at www.hydropower.org. Accessed 22 Oct 2014.

IHA (2013) Activity report: 2013. International Hydropower Association. Available at www.hydropower.org. Accessed 22 Oct 2014.

IRENA (2012a) *Renewable energy technologies: cost analysis series – Hydropower*, Vol. **1** (Power Sector). International Renewable Energy Agency, Issue 3/5.

IRENA (2012b) *Renewable energy technologies: cost analysis series – Hydropower*, Vol. **1** (Power Sector). International Renewable Energy Agency, Issue 5/5.

LeMar P (2002) Integrated energy systems (IES) for buildings: A market assessment. Resource Dynamics Corporation. ORNL/SUB/409200, Final Report.

NREL (2012) Cost and performance data for power generation technologies. Available at http://bv.com/docs/reports-studies/nrel-cost-report.pdf. Accessed 22 Oct 2014.

OECD/IEA (2010) *Energy Technology Perspectives: Scenarios and Strategies to 2050*. International Energy Agency, Paris, France.

Rabiee, A., Khorramdel, H., & Aghaei, J. (2013) A review of energy storage systems in microgrids with wind turbines. *Renewable and Sustainable Energy Reviews*, **18**(0), 316–26.

REN21 (2013) *Renewables 2013: Global Status Report*. REN21, Paris, France. Available at www.ren21.net. Accessed 22 Oct 2014.

Stapleton, G. & Neill, S. (2012) *Grid-connected Solar Electric Systems – The Earthscan Expert Handbook for Planning, Design and Installation*. Routledge, London.

StoRE (2012) Facilitating energy storage to allow high penetration of intermittent renewable energy. D2.1 Report summarizing the current status, role and costs of energy storage technologies. Available at www.store-project.eu. Accessed 22 Oct 2014.

U.S. Environmental Protection Agency (2008) Catalog of CHP technologies. Available at: www.epa.gov/chp/documents/catalog_chptech_full.pdf. Accessed 22 Oct 2014.

Weaver, T. (2012) Financial appraisal of operational offshore wind energy projects. *Renewable and Sustainable Energy Reviews*, **16**(7), 5110–20.

Wiser, R.; Yang, Z.; Hand, M.; Hohmeyer, O.; Infield, D.; Jensen, P.H.; Nikolaev, V.; O'Malley, M.; Sinden, G.; Zervos, A. (2011) Wind energy. *In IPCC Special Report on Renewable Energy Sources and Climate Change Mitigation* [O. Edenhofer, R. Pichs-Madruga, Y. Sokona, K. Seyboth, P. Matschoss, S. Kadner, T. Zwickel, P. Eickemeier, G. Hansen, S. Schlömer, G. von Stechow (eds)]. Cambridge, UK; New York, NY: Cambridge University Press.

3 Modelling Energy Systems

3.1 Introduction

System modelling and simulation is central to the assessment of energy projects. Every energy project must be modelled to some degree of accuracy in order to identify or estimate system size, component types, fuel requirements and energy outputs. Only when parameters and variables such as these have been quantified can the financial and environmental implications of the project and its alternatives be estimated and compared. Often, modelling may not be a recognised explicit step in the project appraisal process, but it is undertaken – whether consciously or not – for all projects nonetheless. For example, the decision whether to upgrade a boiler may involve estimating the quantity and value of gas saved due to the higher efficiency of the replacement and comparing this to the investment required. Here, two systems (the old and the new gas boiler) must be identified, modelled and simulated (albeit very simply) in order to estimate energy inputs and outputs, system sizes as well as capital and running costs.

Energy models may need to be more detailed than this unconscious approach, however, depending on the complexity of the operating environment and the level of accuracy required. For example, the accurate modelling of embedded combined heat and power (CHP) plant subject to time-of-use electrical tariffs with grid export will require the dynamic simulation of heat and power outputs using time steps less than an hour. Energy inputs and outputs are combined with the variable tariffs to estimate cash flows and, ultimately, economic viability. This illustrates the central relationship between energy flows, cash flows and project viability. Inputs and outputs can also be used to estimate emissions and global warming impacts. This is important for assessing the relative environmental impacts of energy projects and influences the design of subsidies, as we will see in Chapter 6.

Renewable Energy and Energy Efficiency: Assessment of Projects and Policies, First Edition.
Aidan Duffy, Martin Rogers and Lacour Ayompe.
© 2015 John Wiley & Sons, Ltd. Published 2015 by John Wiley & Sons, Ltd.
Companion Website: www.wiley.com/go/duffy/renewable

This chapter provides the theory and practical application of energy system identification, modelling and simulation necessary for representing, analysing and optimising energy systems. The approaches outlined can be used to analyse systems using either simple or relatively complex models with energy, emissions and economic inputs and outputs.

3.2   System, model and simulation

System, model and simulation are interrelated concepts, which can be difficult to clearly separate. A system can be thought of as a collection of interrelated objects which is chosen in order to study its behaviour. A model is the representation of this system, to a level of detail which is sufficient to meet the study objectives. Many model concepts exist, but representing systems using mathematical relationships will be the focus of this chapter. It will be seen that different models can be used to represent the same system. Simulation involves running the model to imitate the system's performance over a number of time intervals such as hours, days, months, years or decades. The simulation model can be utilised to analyse the system performance under different operating conditions and configurations. These concepts are described in more detail in following sections.

3.2.1   System

What is a system?

A 'system' is an abstract concept which groups interrelated objects together for the purpose of studying aspects of their behaviour which are of interest. Dandy et al. (2008) define a system as

> a collection of inter-related and interacting components that work together in an organised manner to fulfil a specific purpose or function.

Systems are artificial creations as they isolate and represent specific collections of objects from a much larger set, so that they can be studied for the specific project purpose. This conceptual separation of the chosen system from its surrounding environment is termed the 'system boundary'. A clear boundary must be chosen as any subsequent modelling will otherwise become confusing. It is important to stress that all systems are chosen by humans *for the purposes of the study which they wish to undertake*; that is, the system does not 'exist' in the real world and only exists in the context of the objective for which it has been chosen. It, therefore, follows that the objectives of the relevant study must be clearly understood, so that the system can be defined. For example, a study objective of 'analysing a biomass boiler' is not sufficiently precise to choose a system as a biomass boiler has many performance characteristics: thermal, efficiency, emissions

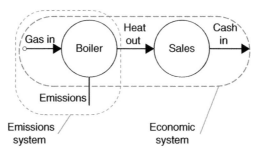

Figure 3.1   Different system boundaries are required for assessing the financial and emissions outputs from a boiler.

or financial. Therefore, a better description might be to 'analyse the financial performance of a biomass boiler'. Further details should be added to better define the system, such as the physical extent of the system or the time horizon of the study. Different systems may need to be chosen for different study objectives for the same collection of objects. For example, analysing the financial and emissions performances of a boiler will require different system representations, as shown in Figure 3.1.

At a high level, a system can be thought of as a 'box' with inputs which are transformed into outputs, but which in every other way is isolated from its surroundings. Inputs are parameters and variables that affect the behaviour of the system, which may or may not be controllable. Outputs are variables which are determined by the input variables and the system processes, which occur within the 'box'. Many sub-processes exist within a system where inputs may be converted to intermediate states, combined, separated, transformed and so on before finally providing the system output. These can be broken down into sub-processes, so that a hierarchal system representation emerges. For example, a wind turbine can be thought of as a single component converting wind energy inputs into electricity outputs, or it can be broken down into multiple sub-components such as the mast, rotor, gearbox, generator and transformers; this abstraction process could be continued theoretically to a molecular or atomic level, although this would be impractical because, at some point, the additional accuracy would not justify the increased complexity. A system can also form part of a super-system when combined with other, related systems. The wind turbine itself forms part of a super-system such as a wind farm and, further, a district and national electrical network (see Figure 3.2).

Some system outputs may form inputs to the system. For example, if we consider a system comprising a heating device and a dwelling, the future temperature of the dwelling is dependent on its past temperature, as well as the energy delivered by the heating device. Therefore, the previous temperature of the dwelling becomes an input to the process of estimating the current temperature (see Figure 3.3). Similarly, in the case of storage systems such as pumped hydro storage, the output from the system affects the stored energy remaining in the reservoir.

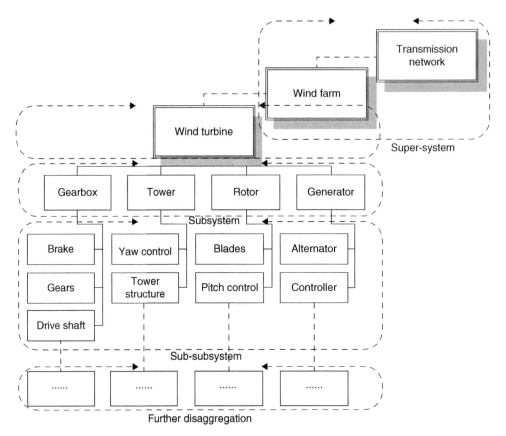

Figure 3.2   Hierarchical system representation.

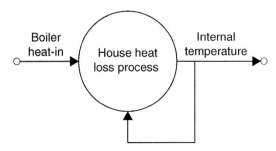

Figure 3.3   Feedback loop in a simple system representing heating in a dwelling. Past temperature outputs influence present and future dwelling temperatures.

System concepts

A system contains *components* (or *entities*) that are the objects and elements which make up the system. For example, components of a geothermal heat pump system might include the ground, ground heat exchanger, heat pump, radiators and hot water tank. The components have data values called *attributes*. The attributes of a ground heat exchanger could be flow rate and temperature. The components have *sub-components* which are systems and sub-systems in their own right. These are organised in a system *structure* which indicates how they are related to one another. This structure is *hierarchical* in that the components are superior to the sub-components and so on. Over time, the system is subject to change as a result of *processes* which normally occur in components. For example, an important process between the ground heat exchanger and the ground is heat transfer. A *controller* is a process which can be used to control the system to meet specified objectives. For example, the heat pump may be switched off when internal temperature set points are met. A system can be thought of as either *static* or *dynamic*. In dynamic systems, system values change over time, whereas in static systems they do not. *System variables* are the system values which change from one time step to another. *System parameters* describe the properties of the system which are typically fixed. For the ground heat exchanger or heat pump system, the flow temperature is a variable, whereas the flow rate may be a parameter if it is fixed on commissioning. The *system state* is the set of variable and parameter values which can be used to completely describe a system at a particular point in time and allows future states to be determined.

Characterising systems

Systems can be either *natural, artificial* or a *hybrid* mixture of both. Natural systems are not created by humans and do not have controllable outputs, whereas artificial systems are designed and created entirely by humans. This book is not concerned with purely natural systems as these do not harness or control energy for human purposes. However, both artificial and mixed systems are of interest as these encompass all of the renewable and energy-efficient systems we will consider. Examples of predominantly artificial energy systems include fossil-fuelled boilers, electrical distribution networks, thermal power stations and many energy storage technologies. None of these rely directly on naturally occurring inputs or processes to produce outputs. Hybrid systems, on the other hand, do rely directly on naturally occurring phenomena to achieve the desired outputs, and they include both energy efficient and renewable energy systems. For example, the in-use thermal performance of an insulated building can only be determined if the artificial system (including walls, windows, insulation, boiler) is extended to include natural phenomena such as external air temperatures, sunshine and wind. Wind turbines are a good example of a hybrid system: the main input is wind speed, which is natural, whereas the artificial turbine sub-system converts this to electricity output. Many renewable energy technologies are hybrid systems because they rely on natural energy sources and artificial energy conversion devices. Fossil- and bio-fuelled systems, however, are artificial systems because both the energy inputs (the fuels) and the energy conversion technologies are typically man-made and controlled (Table 3.1).

Table 3.1   Examples of natural, hybrid and artificial energy systems.

Natural Systems	Artificial Systems	Hybrid Systems
Photosynthesis	Fossil-fuelled devices	Wind turbine
Wind	Bio-fuelled devices	Wave power device
Solar radiation	Thermal power plant	Tidal power device
Water flow	Combined heat and power	Photovoltaic panel
	Pumped hydro storage	Solar thermal collector
	Compressed air storage	Solar power plant
		Hydropower

One difference between artificial and hybrid systems is that the former tend to be more *deterministic* (less random) while the latter are usually more *stochastic* (more random). For example, the amount of electricity generated by a gas-fired power station is almost completely determined by the amount of gas inputted and the efficiency of the system, both of which are known and controlled. By contrast, the energy input to a wave power device is weather-dependent because it varies intermittently and cannot be controlled.

The renewable and energy efficient systems referred to in this book can be classified as *open* systems where both matter and energy are exchanged between the environment and the system. For example, a wood chip boiler uses wood fuel (matter and energy inputs) and generates ash and gaseous emissions (matter outputs) as well as heated water and exhaust gases (energy outputs). This is in contrast to *isolated* systems, which cannot exchange mass or energy, and *closed* systems, which can exchange energy but not mass with the surrounding environment.

3.2.2   Models

A model is a physical or abstract representation of a system. Although the concept is used broadly to encompass physical, mental and verbal models; henceforth, we will discuss mathematical models only as these provide the quantified outputs required for making decisions about energy projects. Traditionally, models were typically physical, often small-scale representations of systems on which experiments were carried out to learn more about the system of interest. Examples include scale models of steam engines, bridges, aircraft wings or room layouts, which were used to analyse process efficiencies, structural performance, aerodynamics or usability. Design and optimisation using this approach, however, can be expensive and time consuming. Such models must be reconfigured physically before a new set of experiments is undertaken and this process may need to be undertaken many times. Although physical models are still used today to validate numerical models or investigate highly complex systems – such as the hydrodynamic behaviour of ships' hulls or coastal engineering projects – the use of mathematical models, which are solved numerically using computers, is now far more common due to their relatively low cost, speed of production, ease of analysis and flexibility.

What is a model?

For the purposes of this book, a model is the mathematical representation of systems for estimating outputs such as energy, emissions and monetary flows from renewable and energy efficient systems. The models described here are 'bottom-up' engineering models of individual energy projects in that the behaviours of individual sub-components are aggregated to replicate the behaviour of the overall system. Important system outputs include the amount of energy produced (normally heat and/or electricity), greenhouse gas emissions and cash flows of costs and revenues which are used to assess financial performances. Cash flows are often largely dependent on energy flows, so modelling one is necessary before modelling the other. Because of this strong link between the energy and economic sub-systems, the process is often referred to a *techno-economic modelling* but is also referred to as *energy analysis* and *energy systems analysis*. Of course, GHG emissions are also linked to energy flows, so the entire process could be referred to as *techno-environmental-economic modelling*, but this is too cumbersome. These relationships are shown in Figure 3.4.

The process of modelling involves representing the chosen system at an appropriate level of detail; it should be sufficiently detailed to achieve the objectives of the study, yet not so complex as to incur unnecessary cost, time or uncertainty (Banks, 1998). Achieving this balance is one of the important skills of modelling. At the beginning of any modelling exercise, there is

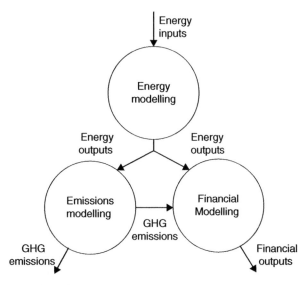

Figure 3.4   Relationships between energy, environmental and economic models of renewable and energy efficient systems.

usually a very large amount of information available, much of which is not relevant to, or is too detailed for the study. The modeller should choose a level of detail based on a clear understanding of the objectives of the study and discard unnecessary information. For example, rather than using half-hourly wind-speed data for the initial modelling of a wind farm, average annual wind speeds and turbine capacity factors may be adequate. Therefore, an important task in modelling is the identification of components, processes and data which are necessary to represent the system for the purposes of the intended study and to separate these from all superfluous information. This process of abstraction and simplification while maintaining the necessary output resolution is a skill developed through practice. The modeller should remember that there is no single correct model which represents a particular system and that typically a number of different models can be successfully employed. Finally, models never truly represent the system being studied and are never exact.

Modelling a system

Modelling a system first involves identifying system inputs and outputs, decomposing the system into suitable individual components, each with its own inputs and outputs, and breaking these down into sub-components until a level of modelling resolution is reached which satisfies the needs of the study. Choosing an appropriate level of model detail is important as it is undesirable to spend excessive time, money and effort in developing the model, nor it is desirable to develop a model which is so simple that it does not represent the system to the level of detail required to achieve the aims of the study. In general, the literature suggests that models tend more towards over-complication than towards excessive simplicity. Bearing this in mind, the modeller should consider the following points when attempting to strike the right balance.

- The measures which will be used to evaluate the project performance should be explicitly identified at the outset of the modelling and simulation process in order to ensure that only the necessary outputs are modelled. For example, it would be a waste of time to model the revenues from a renewable energy supply system if the study objective is to estimate life cycle costs (see Chapter 4).
- The level of model detail should be consistent with the available data. For example, if electricity demand is recorded at hourly intervals, there is little point in building a model with a 15-min time step.
- However, do not be drawn into adding excessive detail just because data or process information is available. Ensure these data are really needed to estimate the desired outputs.
- Because energy is normally difficult to store, the temporal interactions of energy supplies and demands must be carefully considered when choosing an appropriate time step. For example, CHP simultaneously produces heat

and power, but the heat and power demands of a site are rarely synchronised so that valuable energy is wasted. In such a situation, the choice of a suitably small time step can be important to accurate financial modelling.

- The value of energy can vary over short time periods and the model time step must be able to capture these effects. For example, wholesale spot electricity prices vary throughout the trading day, often in hourly- or half-hourly periods.
- Time, cost and resource constraints during the modelling process may affect the level of model detail.

At some stage, adding further costly complexity to the model will only result in small increases in output accuracy and the law of diminishing returns will apply. The skill, therefore, is to know at what level of detail to model. This often involves beginning with the simplest model and progressively adding complexity until the specified or desired level of resolution is achieved or time and resource constraints preclude further development.

Example 3.1   Modelling a System

A facilities manager is investigating whether to upgrade an aging boiler to a modern gas-fired condensing unit. He must estimate gas savings in order to make a financial case to management for the budget necessary for the project. In order to do this, he must develop an energy model of the new system, estimate the amount of gas it will consume and quantify its savings over the existing system. He first sketches the system and maps inputs and outputs (Figure 3.5).

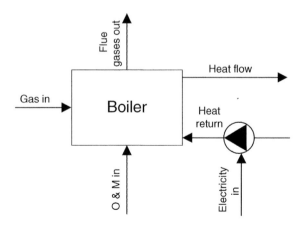

Figure 3.5   Initial sketch of the main boiler system components.

Annual gas consumed is chosen as the measure to evaluate the performance of the proposed system, because he can then easily calculate the amount of gas saved over the existing system and the value of this saving. The system boundary is chosen and excludes: flue gases, because environmental effects are not being considered; pump (and boiler) electrical energy consumption, because these are common to both the proposed; and existing systems and operation and maintenance (O&M) costs, because these will also approximately cancel out for both systems. A 1-year time step is chosen as gas supply and heat demand are exactly coincident (there will be no losses due to temporal supply and demand mismatch) and because the gas tariff is independent of time. The final system model is shown in Figure 3.6. The main input to the system is the building's heat demand, which is related to the amount of gas consumed by boiler efficiency.

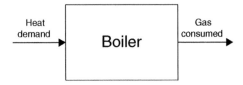

Figure 3.6   Diagram of the boiler system model.

Flow diagrams

It is useful to be able to graphically represent a system model in order to clearly define its boundaries, inputs, outputs, components, sub-systems and behaviours. There is no universally accepted way to represent models: this can be achieved in many different ways depending on the model type and sophistication using process flow charts, flow diagrams, simulation diagrams, functional flow block diagrams or system diagrams. However, the two most important features of any system model diagram are *signals* and *transformations*. A signal represents an input or output value and is represented by a line. A transformation changes a signal from one value to a new value according to an explicit mathematical relationship. The modeller attempts to accurately represent system components, processes, variables, parameters and structure using signals and transformations. This may require some effort, but it is well worth it because it greatly clarifies the modelling process. Flow diagrams will often be reconfigured and refined several times during the modelling process. Very simple diagrams which use signals and transformations only are used here although more detailed representations can be used. Some common signal and transformation combinations used in energy efficient and renewable energy models are illustrated in Table 3.2.

Transformations are typically associated with components or entities. For example, wind speed inputs are mapped to electricity outputs by a wind turbine transformation. However, some transformations are not associated with physical components but with abstract entities. For example, electricity outputs become revenues using a transformation representing energy sales. Inputs are converted to outputs using mathematical transformations, the

Table 3.2   Some common signals/transformation arrangements using simple system diagrams (note that the subscript *t* denotes time).

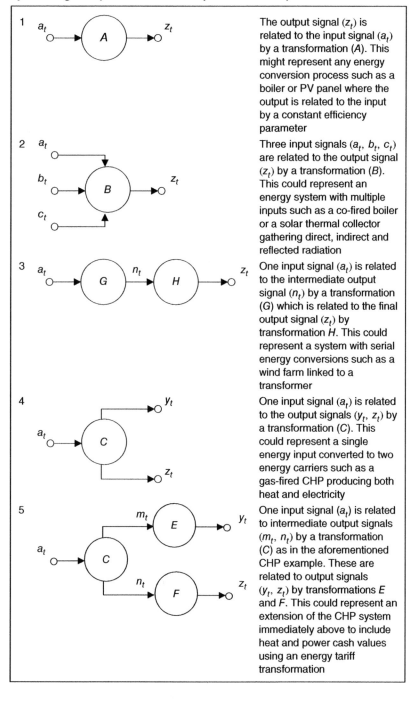

1	The output signal (z_t) is related to the input signal (a_t) by a transformation (A). This might represent any energy conversion process such as a boiler or PV panel where the output is related to the input by a constant efficiency parameter
2	Three input signals (a_t, b_t, c_t) are related to the output signal (z_t) by a transformation (B). This could represent an energy system with multiple inputs such as a co-fired boiler or a solar thermal collector gathering direct, indirect and reflected radiation
3	One input signal (a_t) is related to the intermediate output signal (n_t) by a transformation (G) which is related to the final output signal (z_t) by transformation H. This could represent a system with serial energy conversions such as a wind farm linked to a transformer
4	One input signal (a_t) is related to the output signals (y_t, z_t) by a transformation (C). This could represent a single energy input converted to two energy carriers such as a gas-fired CHP producing both heat and electricity
5	One input signal (a_t) is related to intermediate output signals (m_t, n_t) by a transformation (C) as in the aforementioned CHP example. These are related to output signals (y_t, z_t) by transformations E and F. This could represent an extension of the CHP system immediately above to include heat and power cash values using an energy tariff transformation

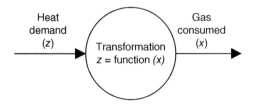

Figure 3.7   A simple model of a boiler system.

nature of which varies depending on the technology or process involved. For example, in its most simplified form, a boiler transforms fuel inputs into heat outputs using an efficiency constant subject to certain constraints. A boiler transformation might be represented by a heat demand (z), an efficiency (η) of 0.7 (i.e. a conversion efficiency of 70%) and a gas consumption (x) as shown in Figure 3.7 and defined in Equation 3.1. If the system input z is given, then we can estimate x for the bounded set of gas inputs: a lower bound may be no gas entering the boiler and the upper bound the manufacturer's maximum rated gas input, x_{max}.

$$x = \frac{z}{0.7}, \quad 0 \le x \le x_{max} \tag{3.1}$$

This very simple relationship is commonly used in techno-economic models for technologies such as CHP, boilers, various generators, pump storage and batteries. The resulting linear relationship between gas input and heat output is shown in Figure 3.8. However, some transformations are more

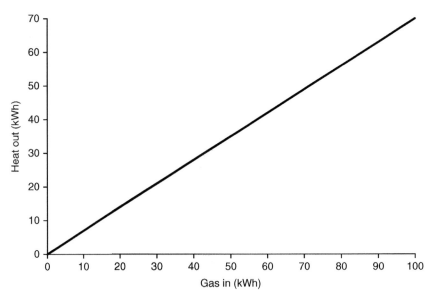

Figure 3.8   A simple relationship between fuel energy input and thermal output for a 70 kW gas boiler (refer to Equation 3.1).

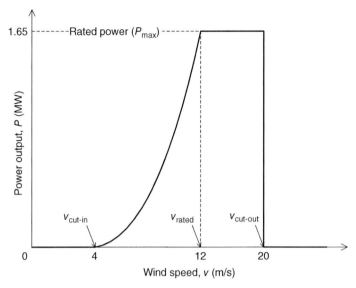

Figure 3.9   An indicative relationship between energy in (wind speed) and out for a 1.65 MW rated wind turbine.

complex. For example, the dynamic output from a wind turbine is represented by a power curve as shown in Figure 3.9 where wind speed and power output are related by Equation 3.2. There are no power outputs for wind speeds less than the cut-in and greater than the cut-out wind speeds ($v_{\text{cut\_in}}$ and $v_{\text{cut\_out}}$, respectively). Above the cut-in wind speed power output is proportionate to the cube of the wind speed up to the rated speed (v_{rated}). Beyond this the unit operates at maximum power output (P_{max}) until it reaches the cut-out wind speed ($v_{\text{cut\_out}}$).

$$P = \begin{cases} 0, \ v_{\text{cut\_in}} \geq v \geq v_{\text{cut\_out}} \\ \frac{1}{2}\rho A v^3 C_{\text{p}}, \ v_{\text{cut\_in}} \leq v \leq v_{\text{rated}} \\ P_{\text{max}}, \ v_{\text{rated}} \leq v \leq v_{\text{cut\_out}} \end{cases} \tag{3.2}$$

where P is the electrical power output from the wind turbine (W), ρ is the density of air (kg/m^3), A is the swept area of the turbine blades (m^2), v is the air velocity (m/s) and C_{p} is the power coefficient of the turbine which varies with the ratio of blade tip speed to wind speed.

This modelling approach can be used for small time steps. However, a simpler relationship for estimating wind energy output over a year is given by Equation 2.8.

Some transformations are control loops where energy output is determined by other system input variables such as energy demand. For example, the electrical or thermal power output from a CHP unit may be configured to meet either the site electrical or thermal demands respectively. In the case where the CHP control strategy is configured to satisfy site electrical demand, the relationship between electrical demand and CHP output could be expressed using an algorithm of the following form:

1. if the site electrical demand is less than or equal to the maximum CHP electrical output, then the electrical output will equal site electrical demand; else
2. if the site electrical demand is greater than the maximum CHP electrical output, then the electrical output will equal maximum CHP electrical output.

This algorithm is easily implementable in spread sheet packages such as MS Excel.

Example 3.2   Developing a Flow Diagram for a Simple System Model

The decision whether to invest in a wind turbine requires the development of a model which will reliably estimate electrical output, O&M costs and sales revenues. The following model is a representation of the system showing transformations in circles (the transformation name is in bold and the mathematical form in italics below) as well as signal connectors and names (Figure 3.10).

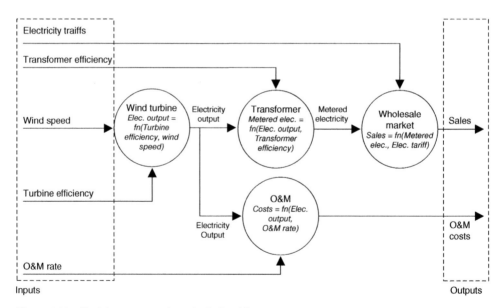

Figure 3.10   Model representation of wind turbine system.

Time

Systems can be represented using either *event-* or *time-driven* models. In the former case the system state only changes if certain conditions are met: an event occurs and the system responds in a certain way. For example, a boiler may not activate unless temperatures fall below a certain level and heat is demanded. In time-driven systems, events are time-dependent: for example, the inputs (solar radiation) and outputs (electricity) to and from a photovoltaics (PV) system are dependent on the time of day. Such systems may either be continuous or discreet. For continuous time-driven systems, conditions are specified for all values of time, which is treated much in the same way as we observe time, as an indivisible continuum. These systems are normally solved analytically. The alternative approach is to use discreet time-steps, where system conditions are calculated at precisely spaced time intervals, say 1 s or a minute apart and the system conditions 'jump' from one state to another. When developing techno-economic models of renewable and energy efficient systems, a discreet time-step approach is normally adopted since continuous data do not exist for all system inputs, many of which are measured in the field at various sampling intervals. Many environmental data (such as wind velocity, air temperature and global solar radiation) are typically collected at time intervals varying from minutes to hours, or even daily or monthly. Critically, the economic performances of real energy systems are normally determined using sampled energy data collected from electricity and heat metres, to which energy tariffs are applied. Time-of-day electricity tariffs are common amongst large users and are being introduced in many European countries at a domestic level. These tend to sample at 15-min, half-hourly or hourly time intervals. In many cases, it may be appropriate to model these systems using similar time steps, although very often much larger monthly or annual time steps are most efficient. Modelling energy systems using a time-step approach typically uses numerical modelling employing algebraic relationships and estimates what has occurred in the intervening time period.

   Take, for example, a simple equation which relates the electrical energy produced by a PV panel (E) to the level of in-plane global irradiation (G) by an efficiency factor η (see Section 2.11). The output can be calculated continuously for all values of time between t and t_0 using the integral in Equation 3.3. However, as irradiation data are collected at discrete time intervals, it is better to model the system using a discreet approach with appropriate time intervals, for example, hourly, to estimate the electrical energy output numerically. The relevant algebraic equation for this approach is shown in Equation 3.4 where G_t represents the irradiation data input value for time interval t. Results are summed over the desired time period using the chosen time step. For example, for a 1-h step, the daily energy output can be calculated by summing the results

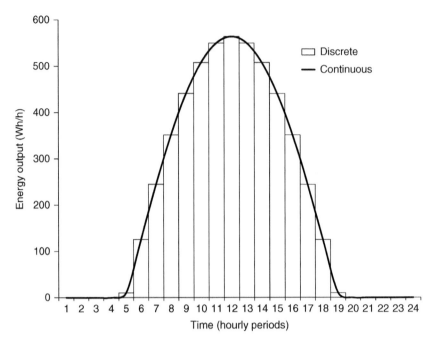

Figure 3.11   A graphical representation of continuous and discrete time–driven representations of a PV system energy output over a 24-h period using Equations 3.3 and 3.4.

for each of the 24-h ($n = 24$) time steps. The two approaches are illustrated graphically over a 24-h period in Figure 3.11.

$$E = \int_{t_0}^{t} G\eta dt, \quad t \geq t_0 \tag{3.3}$$

$$E = \sum_{t=1}^{t=n} G_t \eta \tag{3.4}$$

Analysis

A properly specified model of a system can be used in a number of ways including analysis, design, optimisation, management and control. For example, where both the input and the system behaviour are known, outputs can be calculated; this is referred to as *system analysis*. *System design* concerns determining the components, configuration, sizing and operation of a system which will meet specified requirements. *System management* might involve varying input resources to control outputs in order to meet different environmental conditions. *System control* relates to the design of protocols and devices to regulate system outputs. *System optimisation* concerns varying system design, controls and parameters to maximise a particular output parameter or set of parameters. In this book we are most interested in the interrelated tasks of system analysis, design and optimisation.

Analysis is the task of predicting how an energy system will behave when subject to specific operating and environmental conditions. A wide variety of different analyses can be undertaken, including one or a combination of the following methods. *Sensitivity analysis* assesses the magnitude of changes in system outputs relative to changes in single input variables and parameters. *Scenario analysis* considers plausible future environments by choosing different input value sets, which are internally consistent and assessing their effects on system outputs. *Dynamic analysis* investigates the performance of the system over time. *Steady state analysis* looks at the conditions under which system variables remain constant over time. No system is completely deterministic and *probabilistic analysis* involves the application of statistical methods to quantify the probability of system outputs under different operating conditions. *Monte Carlo analysis* is frequently used in engineering where known or estimated probability distributions of system inputs are repeatedly randomly sampled and output probability distributions for the model are estimated. Although we will not be considering all of these techniques in this book, it is important that the student be aware of the range of analytical approaches available. We will largely illustrate concepts using deterministic, dynamic system models. Some examples of sensitivity analysis and scenario analysis will be given to illustrate how these deterministic models can be extended to better reflect the real, uncertain world in which energy technologies operate.

System design is a complex area which involves identifying alternative design solutions, analysing these solutions and evaluating the extent to which they meet desired outputs. It is typically an iterative process where components and their interrelationships are changed in a series of steps to improve system performance. For example, the performance of a PV system may first be assessed with a single inverter for an array and then with integrated micro-inverters for each module. Once the best inverter configuration option is chosen attention may switch to choosing an optimal module type and array orientation.

Optimisation involves providing the best system performance with respect to the desired objectives. It often involves identifying system inputs and parameters which either maximise or minimise the desired system outputs. In some cases a project may involve choosing between systems with different characteristics such as efficiencies, capital costs or sizes. These system parameters all affect the economic performance of the project and an optimal set of parameters may be chosen, which gives the best possible business case. Tuning part of the system only to optimise a subset of outputs can result in sub-optimal overall system performance and is referred to as the *error of sub-optimisation*. For example, the sizing of a modular solar water heating system for a domestic dwelling is dependent on the demand for hot water. Optimally sizing the collector without considering the size of the storage tank may result in an oversized panel in a system, which performs poorly due to insufficient thermal storage.

Example 3.3   Optimisation of Domestic Energy Efficient Upgrade

A homeowner wishes to insulate her attic and obtains quotes for the investment required for different depths of insulation. She wishes to determine the optimum depth of insulation she should purchase. She decides that a convenient output parameter is the ratio of annual cost savings (over the existing un-insulated baseline) to the cost of the investment – the savings-to-investment ratio. A simple steady-state model is used to represent the system (see Figure 3.12). Heat losses are calculated using heating degree days, the insulation U-value and the insulated area. These are corrected to take account of boiler efficiency and subtracted from the baseline to estimate the value of savings. This is expressed as a fraction of the investment cost to obtain the savings-to-investment ratio (see Section 4.3.3). Once the model is built, it can be experimented on by changing insulation thicknesses to calculate the associated savings and savings-to-investment ratios. The input data and results are shown in Table 3.3. The results can be plotted to identify the optimum insulation thickness based on the savings-to-investment ratio, as shown in the Figure 3.13.

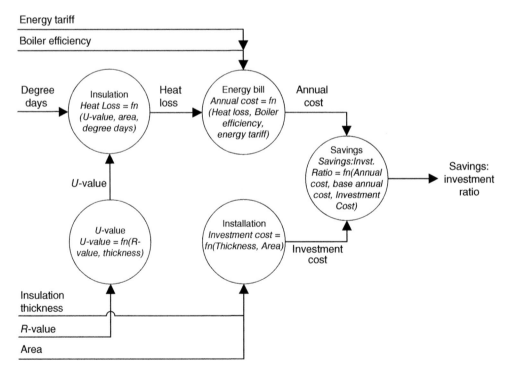

Figure 3.12   Model of attic insulation system.

Table 3.3   Results of the attic insulation model.

Investment cost (€)	Equivalent insulation thickness (mm)	U-value (W/m² K)	Annual heat loss (kWh)	Annual cost (€)	Annual saving (€)	Savings-to-investment ratio (€)
0	25	2.000	2304	329		Base case
700	50	1.000	1152	165	165	0.24
735	75	0.500	576	82	247	0.34
770	100	0.333	384	55	274	0.36
875	125	0.250	288	41	288	0.33
1050	150	0.200	230	33	296	0.28
1190	175	0.167	192	27	302	0.25
1313	200	0.143	165	24	306	0.23
1435	225	0.125	144	21	309	0.22
1575	250	0.111	128	18	311	0.20

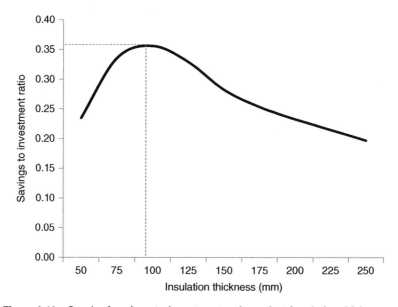

Figure 3.13   Graph of savings-to-investment ratio against insulation thickness showing optimum thickness.

3.2.3   Simulation

Simulation is the process of imitating the operation of a system over a period of time. It can be thought of as automating the system model and launching it through time for multiple time iterations (for discrete time step models). In this way, a synthetic operational history is generated, which can be used for model analysis, design and optimisation, as described earlier. Assessing the economic and environmental performances of renewable energy and energy efficient technologies typically involves dynamic simulation in discrete time steps over an appropriate time period. This is often the lifespan of the investment, but it can also be a period determined by the availability of tariff supports or the investment period stipulated by banks or private investors.

For reasons of speed, ease of analysis and accuracy, computers are normally used to undertake simulation studies. A wide variety of different computer programs can be employed to implement energy model simulations. These can be divided into three main categories:

- *General-purpose application software*, which can be used for a wide variety of tasks, a good example of which is Microsoft Excel;
- *Specific-purpose software applications*, which either are specifically dedicated to the techno-economic analysis of energy supply and energy-efficient systems, or produce energy outputs, which can be easily extended to economic analysis such as EnergyPlus, Homer and RETScreen; and
- *Programming languages* such as C++, MATLAB and Python.

The decision on what software to use to simulate a system model depends on many factors including the complexity of the model, the competencies of the modeller(s) and the available budget and time. Some of these factors are compared in Table 3.4.

An essential task in any modelling activity is to ensure that the resulting model simulation is a sufficiently accurate representation of the system it represents. Confidence is gained through the processes of verification and validation. Verification involves ensuring that model assumptions, transformations, parameter values and the internal relationships have been accurately mapped to the computer simulation program. It is an intrinsic part of the modelling and simulation process. There are a number of approaches to verification:

- ensure that the computer program is reviewed by an independent competent person;
- check that the program responds appropriately to changes in parameter values; and
- compare system state values to those which can be calculated manually.

Validation is the process of establishing whether the simulation model is sufficiently representative of the system for the purposes of the study being undertaken. This is typically a significant task and is only truly possible if data from a version of the system exist against which simulated operational history can be compared. Usually, however, each system is unique and complete verification is not possible. An alternative option is to use data from a similar system and alter the model's parameter values and/or input variables, so that the simulation outputs are as representative of this surrogate system as possible. This approach is most suited to organisations which repeatedly invest in similar energy projects and have these data in-house. For organisations

Table 3.4 Some factors compared when choosing which software application to use for model simulation.

Application type	Model complexity	Analytical complexity	Modeller competency	Time availability	Budget availability
Specific-purpose	Suitable for simple models with generic parameter and data inputs using existing system libraries	Suitable for simple parametric analysis	Suitable for modeller with limited programming experience although training often essential	Usually relatively quick to configure and use	Application cost varies greatly, but labour costs should be minimised using the approach
General-purpose	Suitable for more complex models requiring bespoke programming	Suitable for more complex analysis such as certain type of design and output optimisation	Suitable for those with experience in programming the chosen application	Greater time requirement for programming, verification and validation	Application usually available in-house although labour costs are greater
Programming language	Suitable for complex models capable of handling variables requiring bespoke programming	Suitable for all forms of complex analysis	Suitable for those with the necessary programming skills	Typically requires the greatest time requirement for programming, verification and validation	Application cost varies, but labour costs are usually greatest using this approach

undertaking once-off projects, such data are unlikely to be available and alternative validation approaches are necessary. In such situations only partial validation may be possible. One such approach is to validate sections of the model where relevant data are available. For example, a model of a replacement chiller system could be validated using input and output data for the existing chiller by changing the model parameters to match those of the existing system. A second such approach is to simulate using a different model (possibly a general proprietary software product) and compare outputs.

The measurement of error between simulation model outputs and the reference system observations can be undertaken as outlined below.

- The most basic approach is to compare outputs and observations visually. Small numbers of observation/output pairs can be compared numerically while larger samples can be inspected using scatter or time series plots. However, this does not quantify the accuracy of the model.
- Simple statistical parameters such as means and standard deviations can be computed and compared for both the simulated results and observed data. However, care needs to be taken that these are truly representative of the performance of the system. For example, the means of time series values may be identical, but their temporal representation of the system may be very different.

- Given the fact that most techno-economic system models produce time-series outputs, a better approach is to quantify the difference between outputs and observed data for each time interval. Sample means and standard deviations can then be calculated and a time-dependent measure of model accuracy obtained. Mean percentage error (MPE) is given by

$$\text{MPE} = \frac{1}{n}\sum_{t=1}^{n} \frac{A_t - F_t}{F_t}$$

where A_t is the actual value and F_t is the forecast value and n is the number of time-series data. However, positive and negative errors cancel to give an average error value only. Moreover, MPE is problematic when forecast values are zero, which results in a zero denominator and an undefined result, while small values result in very large errors for similar reasons. The measure is also scale-dependent; for example, different results would be obtained using the celsius and kelvin temperature scales for the same analysis.

- The weakness of positive and negative errors cancelling is overcome by mean absolute percentage error (MAPE). This is the average of the time series of absolute errors between model output-data pairs. It is given by

$$\text{MAPE} = \frac{1}{n}\sum_{t=1}^{n} \left| \frac{A_t - F_t}{F_t} \right|$$

However, the parameter suffers from the same problems of small or zero forecast values and scale effects as MAPE.

- Mean absolute deviation (MAD) overcomes the zero denominator and scale problems of MAPE and is less sensitive to outliers as compared to the standard deviation.

$$\text{MAD} = \frac{1}{n}\sum_{t=1}^{n} |A_t - F_t|$$

However, it does not express the relative size of the error.

- There are a variety of other approaches to estimating model error. These are beyond the scope of this book and further information can be found in Law (2007).

Example 3.4   Error Measurement for a Wind-Farm Model

A wind farm operator owns a wind farm for which monthly energy outputs as well as a wide range of system data such as hub-heights, turbine type, wind speeds and capacity factors are available. It has developed a model of energy output from farm, which it wishes to validate. Using the measured system parameters as inputs, monthly energy outputs are forecast using the model and compared to the actual outputs over the same time period. Table 3.5 gives the actual and forecast (simulated) values. The results are compared using a scatter plot (see Figure 3.14), which shows a high degree of correlation.

Table 3.5   Actual and forecast monthly energy productions (in MWh) for the wind farm showing period (month), mean deviation, mean absolute deviation, mean percentage error and mean absolute percentage error between the two sets of figures.

Period, t	Actual, A_t	Forecast, F_t	Deviation, $A_t - F_t$	Absolute deviation, $\|A_t - F_t\|$	Percentage error, $\frac{A_t - F_t}{F_t}$ (%)	Absolute percentage error, $\left\|\frac{A_t - F_t}{F_t}\right\|$ (%)
1	1204.0	1177.0	27.0	27.0	2.29	2.29
2	1187.0	1213.0	−26.0	26.0	−2.14	2.14
3	1002.0	1062.0	−60.0	60.0	−5.65	5.65
4	1113.0	1105.0	8.0	8.0	0.72	0.72
5	987.0	1001.0	−14.0	14.0	−1.40	1.40
6	842.0	924.0	−82.0	82.0	−8.87	8.87
7	932.0	901.0	31.0	31.0	3.44	3.44
8	1,017.0	998.0	19.0	19.0	1.90	1.90
9	1167.0	1142.0	25.0	25.0	2.19	2.19
10	1274.0	1293.0	−19.0	19.0	−1.47	1.47
11	1312.0	1340.0	−28.0	28.0	−2.09	2.09
12	1295.0	1323.0	−28.0	28.0	−2.12	2.12
Mean	1111.0	1123.3	−12.3	30.6	−1.10	2.86

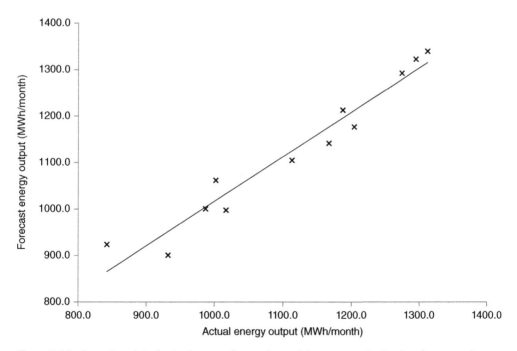

Figure 3.14   A scatter plot of actual versus forecast monthly energy outputs showing a good degree of correlation.

Quantitative measures of error (see Table 3.5) shows the means of the actual and forecast results to be 1111.0 and 1123.3 MWh, respectively, giving a mean deviation of 12.3 MWh. The MAD is 30.6 MWh, while the MPE and MAPE are −1.10% and 2.86%, respectively. The low MPE shows the model predicted energy production with a high degree of accuracy, while the low MAPE indicates that there are no large errors within individual time intervals, which might cancel one another.

Representative outputs are dependent not just on a well-designed model, but also on the quality of the data inputs. It is, therefore, important that the modeller be cognisant of potential data errors and chooses representative, high quality data for inputting to the model. There are many potential data problems to consider, some of which are listed as follows.

- Data may not be representative of the system being considered. For example, 1 year of hourly wind data may not be appropriate for modelling life cycle power output from a wind turbine because they may not be representative of typical long-term wind conditions at the site in question. For this reason, data for a typical meteorological year (TMY) or a longer time series are often used.
- They may not be in the correct format. For example, when modelling PV systems, solar energy data are typically in the form of insolation incident on a horizontal surface which must be transformed to the plane of the PV panel before being used.
- They may contain significant measurement error. For terrestrial solar irradiance measurements significant error can occur due to poor maintenance of measurement devices.
- They may be biased. For example, returns of self-reported household energy consumption may be biased towards owner-occupiers because tenants may not be fully responsible for their energy bills. Data from suppliers and manufacturers on efficiencies, reliability and costs should be treated with caution because they may be over-optimistic.

3.3 Modelling and simulating energy systems

3.3.1 Steps in simulating energy projects

There are many approaches to developing and simulating models to represent renewable and energy efficient systems. In general, modelling and simulation is an iterative exercise, which should incorporate verification and validation as integral steps in the process. Figure 3.15 illustrates the main steps in one approach to modelling and simulating renewable and energy efficient systems.

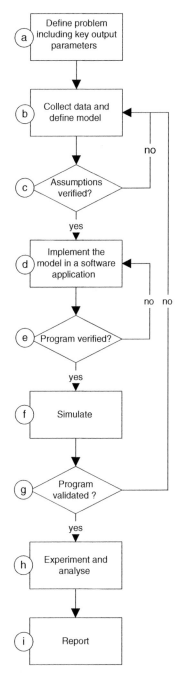

Figure 3.15   The main steps in an approach to modelling and simulating an energy system (adapted from Law, 2007).

Problem definition

The objectives of the study are agreed by relevant stakeholders and clearly defined in quantitative terms. The scope of the system to be modelled should be identified and the performance measures, which will be used to evaluate it specified.

Model building

This involves identifying the signals and transformations to use and gathering the necessary input data for the model. An appropriate level of model detail is chosen to represent the system for the purposes of the study which depends on the nature of the project performance evaluation measures, format and availability of relevant data; importance of temporal effects and project resource constraints (see section on 'Modelling a System'). Remember not to over-complicate the model and that it need not directly represent every system component. It is important to log any assumptions made during this process for verification and subsequent model improvement. Data from existing or similar systems should be collected during this phase for undertaking model validation at a later stage.

Verify model assumptions

It is useful to formally present the model to experts (possibly colleagues or other knowledgeable individuals working in the relevant technology area) outlining the assumptions being made and seeking critical feedback before the model is implemented as a program.

Software implementation

A software application or programming language is chosen depending on factors such as software cost, the competencies of the modeller, the time available, the availability of suitable specific application software and the complexity of the model.

Program verification

Once the program is developed, it is verified by debugging, so that the program matches the model specification. This is described in Section 3.2.3.

Simulation

The program can now be used to simulate the system's performance over the desired time period. Initially, data inputs are chosen to match the known inputs of the system, similar systems or system components (if any are available) for the purposes of program validation.

Program validation

The program outputs and system states are compared to data from the existing system, similar systems or similar system components in order to fully or partially validate the program (see Section 3.2.3). Even if there are no suitable data available for validation, the program should be tested to ensure outputs and system variables behave correctly when input parameters are varied. A sensitivity analysis can be used to identify the most important transformations and parameters (although this may often be obvious), so that these can be given a higher level of attention as compared to less important ones. Analysis can commence when the program is successfully validated.

Experiment and analyse

The model can now be used to achieve the aims and objectives of the study whether that be general analysis, design, optimisation, management or control. This involves designing an experiment to investigate configurations and input parameters of interest and recording the key output parameters originally chosen to assess project performance.

Report

The report should document the modelling process including a detailed description of assumptions, the program and code or application as well as verification and validation results. The results and conclusions of the study should be presented.

3.3.2   Simulation tools

A summary of selected proprietary computer simulation programs relevant to a variety of energy efficiency and energy supply analyses is given in Table 3.6.

3.3.3   Data sources

There are a wide variety of data sources used in the techno-economic modelling of energy systems. The data employed depend on the specific system, geographic location, available technologies and market conditions among other factors. The very limited sources listed subsequently should be treated only as an indication of the scope and typical repositories of data internationally. In general, there are a wide variety of data available from national statistics' offices, government departments and energy agencies. For example, a wide range of energy statistics are available from the UK's Department of Energy and Climate Change (DECC), the French Ministère de l'Écologie,

Table 3.6 Summary of a variety of energy-related simulation tools.

Software	Technologies	Strengths	Weaknesses	Website
AEPS System Planning	Electrical power systems	Uses high level or detailed system model inputs. Provides detailed results		http://www.alteps.com
BLCC	Buildings	User-friendly. Performs high quality, detailed LCC analyses quickly and accurately	User related information for improvements	http://www1.eere.energy.gov/femp /information/download_blcc.html
BEES	Buildings	Offers a blend of environmental science, decision science and economics. It is based on consensus standards, and is designed to be practical, flexible, and transparent	Includes limited environmental and economic performance data	http://www.bfrl.nist.gov/oae /bees.html
BuildingAdvice	Commercial buildings	BuildingAdvice provides fast, easy, low cost energy efficiency analyses of commercial buildings, up to an ASHRAE Level II audit standard	BuildAdvice currently is not designed to perform ASHRAE Level-3 audits of specific HVAC equipment	http://www.airadvice.com
CHP Capacity Optimiser	CHP	The tool provides an efficient means to determine the capacities of prime mover and absorption chiller that maximise economic benefit on a life-cycle basis	Multi-unit prime movers limited to two identical units at this time. No equipment libraries available	http://www.ornl.gov/sci /engineering_science_technology /cooling_heating_power/
EnergyPerisope	Solar and renewable	Develops reports for single- or multi-technology energy solutions. Helps maximise investment efficiency	Not an engineering tool. Used mostly by sales and marketing personnel	http://www.energyperiscope.com/
Homer	Microgrid including wind, solar and storage	Compares different technologies including hybrids. Considers storage. Considers seasonal and daily variations in loads and resources. Designed as optimization model for sensitivity analyses	Does not consider intra-hour variability and variations in bus voltage	http://www.nrel.gov/homer

Name	Technology	Description	Limitations	Website
Polysun	PV, solar thermal, air-ground-source and water heat pumps	Has a large customizable component catalogue. Good for solar thermal and heat pump simulation	Limited to grid-tie PV systems	http://www.solarconsulting.us/polysun.html
PV*SOL	Photovoltaics	Precise calculations possible and very user-friendly		http://www.valentin.de
PVSYST	Photovoltaics	Very refined hourly simulation, large component and meteo database		http://www.pvsyst.com
RETScreen	Photovoltaics, solar thermal, biomass, heat recovery	Easy completion of feasibility studies for renewable energy and energy efficient technologies		http://www.retscreen.net/
T*SOL	Solar thermal	Precise calculations possible and very user-friendly		http://www.valentin.de
Tetti FV	Embedded PV	Large component database, easily extendable		http://www.studioiesl.com
TRNSYS	Photovoltaics, wind, solar thermal, biomass, heat pumps, heat recovery	Modular and flexible for modelling a variety of energy systems. Extensive documentation available. Graphical interface with drag and drop components. Extensive libraries available	No assumptions about the building or system are made. The user needs to have detailed information about the building or system.	http://trnsys.com
Umberto	Life cycle energy, environment, cost	The modelling of process systems is not limited to a specific sector or field of application. Usable in many practical applications in industry, research and consulting	Not applicable as operational support, but rather as a planning or controlling instrument	http://www.umberto.de/en
UrbaWind	Wind	Automatic CFD software for wind modelling in the built environment. Contains a robust solver	Cannot be used for calculating wind extrement pressure on buildings for building construction. No thermal calculations are included	http://www.meteodyn.com

du Développement Durable et de l'Énergie. Data on energy conversion, energy end use and efficient technologies are available from manufacturers although these should always be treated with caution. Eurostat, the US Department of the Environment and the International Energy Agency all publish a wide variety of comparative national and international energy-related data.

Fuels

Data on the energy densities, emissions factors and conversion factors for fuels are available from many sources. The US Department of the Environment publishes 'Unit Conversions, Emissions Factors, and Other Reference Data', which gives a wide variety of data on many different fuels.

Meteorological

Meteorological data are typically available from national meteorological offices, which give daily or hourly data which are important for modelling renewable energy system outputs. Wind, tidal and wave resource maps are available for many countries, especially those with a policy focus in the area. For example, in Ireland, the Sustainable Energy Authority of Ireland produces maps of wind, tidal and wave resources for the country. In the United States, the National Renewable Energy Laboratory maintains a wind resource map. A number of freely available solar resource maps such as SolarGIS can be used in modelling solar technologies.

Energy conversion and end use

Energy conversion is a broad area involving multiple technology types. In many cases, the performance data on important parameters such as efficiencies are only available from manufacturers and are, therefore, not independently verified. Some data repositories of performance data for different manufacturers are available. For example, in the United Kingdom, Seasonal Efficiency of Domestic Boilers in the United Kingdom (SEDBUK) provides data on seasonally adjusted efficiencies for domestic boilers.

Prices

Energy price indices are published by many countries for a variety of fuels through national statistics offices or other government agencies. Some countries such as Ireland (eirgrid.ie) and Denmark (energinet.dk) provide hourly wholesale electricity price data. Energy price projections can be found in the reports of international agencies.

Sector specific

There many reports which contain useful sector-specific data. The US National Laboratory publishes the Transportation Energy Data Book. Many countries provide data on energy use from buildings. In some countries sample high time resolution (e.g. half hourly) from electricity and gas smart metering trials are available at an individual building level.

3.4   Case studies

The following case studies illustrate the application of the modelling and simulation techniques described in this chapter. Different time steps and levels of modelling detail are used in the samples depending on the project objectives and data availability. In all cases energy, cash and emissions flows are simulated using Microsoft Excel. The example Excel workbooks can be found at <www.wiley.com/go/duffy/renewable>. The layout of each example generally follows the steps shown in Figure 3.15.

3.4.1   Office PV system

Problem definition

An small office owner is considering installing a PV system on their building to generate on-site electricity, thus displacing electricity currently imported from the national grid. Any units of electricity not used on site are exported (or 'spilled') onto the grid and attract an export feed-in-tariff from the local energy supply company. The owner employs a consulting energy analyst to estimate the cash inflows for the system. This will enable the owner to undertake further financial analysis.

The physical scope of the system includes all components and work involved in retrofitting the PV system to the building. This comprises the PV array and mounting system, inverter, controller, switch fuse and wiring between the PV array and the main electrical switch room. Inputs to the system which are identified include:

- the electrical demand of the office building;
- the solar energy available to the system;
- component efficiencies, lifespan, degradation factors;
- energy tariffs for avoided and spill electricity; and
- maintenance costs.

The owner specifies that the system performance variable to be evaluated during the modelling study is the value of cash inflows arising from avoided imported electricity and any excess electricity exported onto the grid.

Model building and verification

The first step involves identifying all transformations and signals, so that a representative model can be developed. A wide variety of data are obtained for the project. The building has a time-of-day meter, which records electrical energy consumption at half-hourly time intervals. Hourly in-plane (at right angles to the PV array) solar radiation data are obtained for the nearest meteorological station. System performance parameters are obtained from the preferred PV installation contractor including array and inverter efficiencies. Energy tariffs are obtained from the electricity supply company.

Given that electricity output from the PV system is dependent on the levels of solar insolation, there will be a mismatch between PV output and office electricity demand, so that electricity will be either exported or imported almost continuously. The energy supply company calculates import and export billing using half-hourly meter readings so a half-hourly simulation time interval is chosen as the most representative time-step. However, solar radiation data are only available at hourly intervals, so linear interpolation is used to estimate a half-hourly input radiation dataset.

Important assumptions made during the modelling and simulation process are recorded during the modelling process. They include:

- the office's electricity demand profile for the previous year is representative of those for the lifetime of the PV system;
- operation and maintenance costs are ignored in this part of the modelling process and will be considered in more detail during financial appraisal;
- although system performance degrades slightly each year, this effect is ignored;
- inverter efficiency is constant over its input range;
- a 100-kW_p-sized array is modelled on the basis of client preferences; and
- any wind and temperature effects on PV efficiency are ignored.

The parameter values used in the calculations are shown in Table 3.8. Table 3.7 shows the transformations used in the model to estimate system cash flow outputs for each half-hourly time step. The model is shown in Figure 3.16.

Software implementation, verification, simulation and validation

The model is verified with other energy consultants, by a formal review by a second consultant and through discussions with industry and supply specialists. The model is then implemented in Microsoft Excel using the transformation formulae shown in Table 3.7 as well as the data and parameters described in Table 3.8. It is simulated using half-hourly time steps over a 1 year period using metered time-of-use demand and interpolated solar radiation data inputs.

The program is verified by a second consultant who checks cell formulae against transformations. The model is partially validated using a commercially available proprietary lumped annual electricity output model for the PV system in the same location. PV model output data are summed over one year and compared to the outputs from the proprietary model. Further verification

Table 3.7 Relationships between inputs and outputs for PV system model transformations.

Transformation		Nomenclature	Explanation
Name	**Equation**		
PV array	$E_{DC,t} = G_t \times A \times t \times \eta_{PV}/1000$	$E_{DC,t}$ is the amount of direct current electrical energy produced in time interval t (kWh); G_t is the insolation (W/m^2); A is the array area (m^2); and T is the simulation time step expressed as an hourly fraction; η_{PV} is the PV module efficiency	Sunlight is converted to DC electricity
Inverter	$E_{AC,t} = E_{DC,t} \times \eta_{inv}$	$E_{AC,t}$ is the amount of alternating current electrical energy output in time interval t (kWh). η_{inv} is the inverter efficiency	DC is converted to AC electricity
Energy balance	$E_{av,t} = E_{AC,t,}$, if $E_{dem,t} > E_{AC,t,}$ $E_{av,t} = \begin{cases} E_{AC,t}, & \text{if } E_{dem,t} > E_{av,t} \\ E_{dem,t}, & \text{if } E_{dem,t} < E_{av,t} \end{cases}$ $E_{ex,t} = E_{AC,t} - E_{dem,t,}$, if $E_{AC,t,} > E_{dem,t}$	$E_{av,t}$ is the avoided import electricity in time interval t (kWh); $E_{dem,t}$ is the household electrical demand (kWh); $E_{ex,t}$ is the exported electricity	The model checks if AC electricity output or office demand is greater and calculates the avoided electricity and electricity exported to the grid
Revenues	$F_t = \begin{cases} E_{av,t} \times Ta_{im}, & \text{if } E_{av,t} > 0 \\ E_{ex,t} \times Ta_{ex}, & \text{if } E_{ex,t} > 0 \end{cases}$	F_t is the net cash flow from the PV system in time interval t(€); Ta_{im} is the import tariff (€/kWh); Ta_{ex} is the export tariff (€/kWh)	The value of electricity export or import is determined using the appropriate tariff and a maintenance charge is added

is undertaken by varying input parameters to ensure that the model outputs respond as expected thus indicating that it is well behaved.

A 1-year simulation of half-hourly cash flows is then carried out to fulfil the requirements of the study. A 1-day sample of selected intermediate values is shown in Figure 3.17, while the final output cash flows for corresponding periods are shown in Figure 3.18. Table 3.9 gives annual quantities and values of electricity consumed by the office as well as that produced, exported and avoided by the PV system.

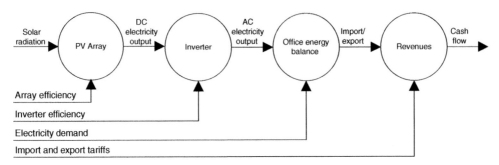

Figure 3.16   Model of the PV system showing main signals and transformation.

Table 3.8   PV system model parameter values.

Parameter	Symbol	Value
Electricity import tariff	Ta_{im}	0.2 €/kWh
Electricity export tariff	Ta_{ex}	0.4 €/kWh
PV area	A	100 m^2
PV module efficiency	η_{PV}	17.2%
Inverter efficiency	η_{inv}	90%

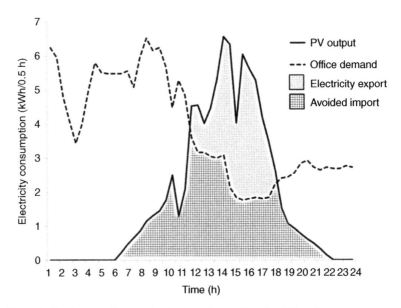

Figure 3.17   Intermediate and output system values for PV system.

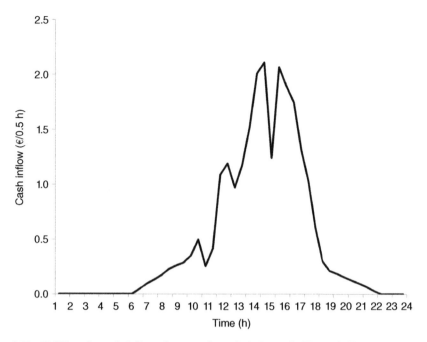

Figure 3.18   Half-hourly cash inflows for sample period shown in Figure 3.17.

Table 3.9   Annual quantities and values of electricity consumed by the office as well as produced, exported and avoided by the PV system.

Parameter	Quantity (kWh)	Value (€)
Total annual electricity use	103,557	20,711
Avoided electricity imports	14,390	2,878
Electricity exports	1,873	749
Total PV electrical output	16,263	3,627

3.4.2   Gas heat pump for data room cooling

Problem definition

The facilities manager of an office building investigates the alternative cooling systems for the data room which houses central Information and Communications Technology (ICT) facilities. The heat produced by the ICT equipment must be controlled such that the ambient temperature of the room is maintained in the range of 18–27°C. A number of cooling alternatives exist, which must be assessed and compared, one of which is a gas heat pump (GHP) producing chilled water to cool the space. A GHP uses natural or liquid petroleum gas (LPG) to fuel an engine which drives a compressor that removes heat from chilled water circuit and 'dumps' it into the external environment. This can be more efficient as compared to electricity because the inefficiencies of electrical energy conversion and transmission processes are avoided. An additional benefit is that some of the rejected heat can be used in the building's hot water system.

The purpose of the study is to identify the typical annual cash flows associated with the GHP investment, so that it can be compared to the other chiller technologies.

The data room has a constant 250 kW thermal load, which will be met by five GHPs each with a rated 50 kW output. Because the unit will connect into an existing pumped-chilled water ring main, neither pipework nor pumping costs are considered. There is an existing gas supply on site. Therefore, the physical scope of the system extends from the gas input to the chilled and hot water outputs.

Model building and verification

Data are first gathered which can be used to model the system. These include energy demands, device efficiencies, heat outputs, energy tariffs and O&M tariffs. The main variable input is chilled water demand as well as parameters such as tariffs and the coefficient of performance (COP). The main outputs include chilled and hot water, the latter being treated as a revenue because it is a beneficial by-product, which avoids the cost of producing some hot water using gas-fired boilers. The capital cost of connecting the hot water output to the building's heating system must, however, be included in any subsequent financial study. The main signals and transformations are illustrated in Figure 3.19 and are listed in Tables 3.10 and 3.11.

The main input parameters – chilled water demand, gas tariffs, heat tariffs and O&M costs are constant. However, the COP of the GHP varies with external ambient temperature, but this effect can be approximated over time using a seasonally averaged COP value. When choosing the number of time steps for the simulation period, it is important to know whether a large time step gives

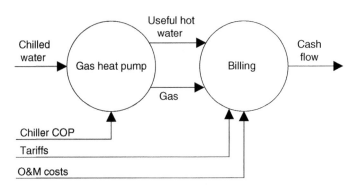

Figure 3.19   Model of the GHP system showing main signals and transformations.

Table 3.10 Relationships between inputs and outputs for GHP system model transformations.

Transformation		Nomenclature	Explanation
Name	**Equation**		
Gas heat pump	$E_g = E_{cw}/COP$ $E_h = E_{cw} \times HCR$	E_g is the amount of gas energy used (kWh); E_{cw} is the amount of chilled water energy demanded in one year (kWh); COP is the seasonal coefficient of performance of the GHP; E_h is the amount of heat energy produced (kWh); HCR is the heat to chilled water produced ratio	The gas burned by the GHP produces a COP number of units of chilled water. Useful hot water output is related to chilled water production by the HCR
Billing	$F = E_h \times Ta_h - E_g \times Ta_g$ $- E_{cw} \times Ta_{main}$	F is the cash flow over the year (€); Ta_g the gas tariff (€/kWh); Ta_h the heat tariff (€/kWh); Ta_{main} is the O&M tariff (€/kWh chilled water)	Cash flow is the sum of the cost of gas, avoided heat and the cost of maintenance

Table 3.11 GHP system model parameter values.

Parameter	Symbol	Value
Avoided heat tariff	Ta_h	0.06 €/kWh
Gas tariff	Ta_g	0.05 €/kWh
Seasonal coefficient of performance	COP	1.4
Chilled water energy requirement	E_{cw}	2.19×10^6 kWh/annum
Hot water-to-chilled water ratio	HCR	0.357
Operation and maintenance tariff	T_{main}	0.0023 €/kWh$_{cw}$

similar output results to a smaller one summed over the same time period. In such circumstances, the larger step is preferable for reasons of efficiency. As all of the inputs can be represented as time-independent parameters, the maximum time interval of 1 year (specified in the study objectives) is chosen. The cooling requirements of the data room are, therefore, estimated using a steady-state model.

Software implementation, verification, simulation and validation

The model signals, transformations and input parameters were verified by a colleague with experience of chilled water systems and the response of the output behaviour for a range of input parameters was checked for consistency. These parameters and transformations were implemented in Microsoft Excel and outputs for one year of operation were estimated. Results are shown in Table 3.12. The system is projected to cost €36,341 to operate each year.

Table 3.12   Outputs for the gas heat pump simulation.

Chilled water output (E_{cw}) (kWh/annum)	Gas input (E_g) (kWh/annum)	Hot water output (E_h) (kWh/annum)	Cash flow (V) (€/annum)
2,190,000	1,564,286	781,830	−36,341

3.4.3   Compressed air energy storage

Problem definition

An energy supply company is considering investing in a compressed air energy storage (CAES) plant with a rated electrical output of 300 MW, which is connected to the electrical transmission network. It stores electrical energy when unit prices are low and supplies electricity when prices are high. Typically, spot market prices, known as system marginal prices (SMPs), are lower at night and higher during the day, thus offering opportunities for arbitrage. As an initial assessment, the company wishes to assess the revenues the plant would have earned over the previous year using publically available half-hourly SMPs. To simplify this initial assessment they assume the new plant does not affect the market price of electricity.

The purpose of the study, therefore, is to estimate the net cash flows from the plant as if it had operated over the previous year. For the purposes of the study, net cash flows are the difference between energy sales and O&M costs. Results are required for a range of different daily compression and generation schedules.

The CAES plant uses relatively cheap off-peak electricity to drive compressors and store pressurised air in the ground. During peak periods when SMPs are higher, the compressed air is released and heated rapidly by burning it with natural gas to drive turbines and produce electricity. Compression costs and energy sales depend on the SMP for each time period. Gas and O&M costs are fixed at a rate per unit of gas consumption and electrical output, respectively.

Model building and verification

A variety of data are gathered for the project. Half-hourly SMPs for the previous year for the market in which the CAES plant would operate are obtained from the grid operator. Transmission use of system (TUoS) charges of 4 €/MWh are added to these according to published rates from the grid operator to estimate the tariff for electricity purchased for compression; unit electricity sale prices are at the SMP without TUoS. Gas supply tariffs are obtained from the wholesale gas market and include transmission and distribution costs. O&M tariffs and system performance characteristics are obtained from the manufacturer and verified with various published reports.

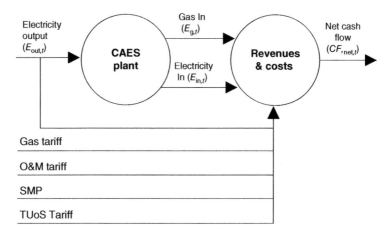

Figure 3.20   Model of the CAES system showing main signals and transformations.

A lumped (or aggregated) model of the overall CAES system is used because performance parameters are available for this level of model detail and because the purpose of the model is to assess the overall energy and cash flows; it is not intended for design assessment or technical optimisation. A model of the CAES plant showing transformations and signals is developed and is shown in Figure 3.20.

The assumptions made during the modelling and simulation process are logged and are summarised subsequently. It can be seen that a comprehensive cash flow analysis of the CAES plant would necessitate significantly more detailed techno-economic modelling which addresses many of the assumptions made.

- Only the value of energy outputs from the CAES plant is considered. Other revenue streams from 'ancillary services' or 'system services' such as capacity, operating reserve, ramping or black start are, although these are likely to be significant.
- Compression cycle electricity is purchased at the SMP plus a TUoS charge of 4 €/MWh, whereas all electricity generated is sold at the SMP.
- System efficiency parameters remain unchanged for the duration of each compression and generation cycle.
- The total stored energy (1200 MWh) is completely discharged each day in one continuous 4-h period cycle by the 300 MW plant.
- The total pressure energy (1200 MWh) is restored each day in one continuous 6-h period by the 235 MW compression train operating continuously at 85% efficiency.
- System generation and compression scheduling is for the same daily periods for each day of the year. As SMPs at weekends are normally quite

different from those on weekdays, different weekday and weekend schedules should be considered; however, these are not simulated.

● Ramp up and down rates are ignored and the system is modelled as if it instantaneously switches on and off.

Given that SMPs vary half-hourly and directly affect cash flows, a half-hourly time interval was chosen for simulation purposes. Tables 3.13 and 3.14 show the transformations and parameter inputs, respectively, for the CAES model. It may seem strange that electricity output is an input to the CAES plant, but for the purposes of modelling, Gas and electricity in are functions of the scheduled output from the plant, thus it constitutes an important model input.

Software implementation, verification, simulation and validation

The model was implemented in Microsoft Excel and net cash flows were calculated using SMPs for the previous year. The model structure, transformations and main input parameters were verified by checking with colleagues,

Table 3.13 Relationships between inputs and outputs for the CAES model transformations.

	Transformation	Nomenclature	Explanation
Name	Equation		
CAES plant	$E_{g,t} = E_{out,t} \times \eta_{HR}/3.6,$ for $t_{gen,on} \leq t \leq t_{gen,off}$ $E_{in,t} = E_{out,t}/\eta_{comp},$ if $t_{comp,on} \leq t \leq t_{comp,off}$	$E_{g,t}$ is the quantity of gas used by the CAES plant in time interval t (MWh); $E_{out,t}$ is CAES electrical energy output in time interval t (MWh); η_{HR} is the heat rate; $t_{gen,on}$ is the scheduled generation on-time; $t_{gen,off}$ is the scheduled generation off-time; $E_{in,t}$ is the compressor electrical input in time interval t (MWh); η_{comp} is the compressor efficiency; $t_{comp,on}$ and $t_{comp,off}$ are the scheduled compressor on and off times, respectively	Periods of generation and storage (compression) are scheduled by the modeller. Gas consumption is determined as the product of the plant heat rate and electricity output. This is divided by 3.6 to convert heat rate from GJ/MWh to MWh/MWh. Compressor electricity consumption is determined as the product of the output and the plant energy ratio
Revenues and costs	$F_{net,t} = Ta_{sys,t}E_{out,t} - Ta_{g}E_{g,t}$ $- (Ta_{sys,t} + Ta_{TUoS})E_{in,t}$ $- Ta_{main}E_{out,t}$	$F_{net,t}$ is net cash flow in period t; $Ta_{sys,t}$ is the system marginal price for period t (€/MWh); Ta_{g} is the gas tariff (€/MWh); Ta_{TUoS} is the TUoS charge (€/MWh); Ta_{main} is the maintenance cost (€/MWh)	Net cash flow is the energy output revenues less costs of gas, compressor electricity and maintenance

Table 3.14   Parameter inputs for the CAES model.

Parameter	Symbol	Value
Gas tariff	T_g	13 €/MWh
Heat rate	η_{HR}	4.2 MJ/MWh
Compressor efficiency	η_{comp}	85%
O&M tariff	T_{main}	2.1 €/MWh
Transmission use of system charge	T_{TUoS}	3 €/MWh
CAES plant electrical output capacity	P_{out}	300 MW
Compressor electrical rating	P_{comp}	235 MW
Useful energy stored in CAES	E_{sto}	1200 MW
Compression period	E_{sto}/P_{comp}	6 h
Generation period	E_{sto}/P_{out}	4 h

Table 3.15   Simulation cumulative annual net cash flow results (in €m) for different compression and generation scheduled on times.

		Generation on time			
		00:00	06:00	12:00	18:00
Compression on time	00:00	−11.2	−7.7	0.3	7.1
	06:00	−18.8	−15.3	−7.4	−1.2
	12:00	−26.7	−23.2	−15.2	−8.4
	18:00	−28.7	−25.2	−17.2	−10.4

literature and industry experts. A sensitivity analysis was undertaken to ensure that changes in simulation outputs were consistent with changes in parameter inputs. Table 3.15 shows a matrix of cumulative annual net cash flow outputs for selected compression and generation schedules. It can be seen that the highest simulated cumulative annual net cash flows were obtained when compression and generation occurred during off-peak and peak hours, respectively. Conversely, lowest net cash flows were estimated for generation during off-peak and compression during peak periods. Figure 3.21 shows one pair of compression and generation schedules for a single day and illustrates why the highest cash flows occur when generation and compression periods are chosen to coincide with high and low SMPs, respectively.

3.5   Conclusions

Techno-economic modelling of energy systems is central to cash flow estimation. The amount of time and effort put into developing models, gathering data, validating and verifying and interrogating the model will be reflected in better cash flow estimation and, consequently, better financial decision making. Modelling is often not undertaken as a formal step in the cash-flow

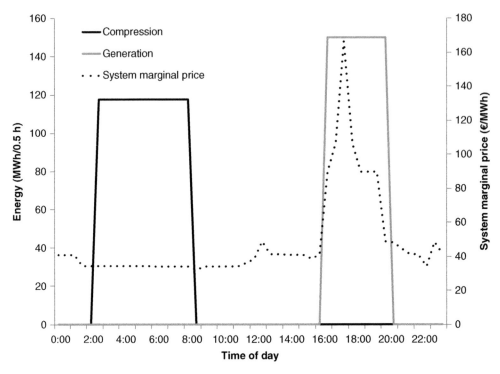

Figure 3.21   Graph showing CAES compression and generation schedules and electricity system marginal price (SMP) for a 1-day period.

estimation process, but this is unwise because the process of formally defining system boundaries, choosing an appropriate level of modelling detail and time step, interrogating data inputs as well as verifying and validating simulation outputs helps provide clarity to the project appraisal process and ultimately results in time savings and better results, even for small projects. It is important to question the data inputs because if these are inaccurate, even the best model transformations will yield inaccurate results. The modeller should constantly critically analyse the data sources: who is the provider and do they have an incentive to report very optimistic or pessimistic figures? Are they representative of the project under analysis? What data gathering techniques were used and are they reliable? The degree of rigour applied to the modelling process should be appropriate to the complexity and importance of the project. For example, modelling a large grid energy storage project may take many months or years, involve many primary and secondary data sources and require many iterations before decision makers are satisfied that modelling risk has been reduced to an acceptable level. Although many proprietary software applications are available and suitable, simple spreadsheet models can be used for many projects, up to and including very complex ones. Finally, a good model which has been suitably validated is a very useful tool, which can be used to simulate and optimise a wide variety of different project scenarios and add value to an organisation, individual or society.

References

Banks, J. (ed) (1998) *Handbook of Simulation: Principles, Methodology, Advances, Applications and Practice*. Engineering and Management Press; John Wiley and Sons, Inc, New York.

Dandy, G., Walker, D., Daniell, T., and Warner, R. (2008) *Planning and Design of Engineering Systems* (2nd ed.). Taylor and Francis, New York.

Law, A. (2007) *Simulation, Modeling and Analysis* (4th ed.). McGraw Hill, New York.

4 Financial Analysis

4.1 Introduction

Financial analysis, also referred to as investment appraisal or capital budgeting, is the process of establishing the economic value of investments which typically include high up-front costs, are long-lived and are of strategic importance to their investors. Many energy projects are capital-intensive and long-term, and the decision to invest requires the commitment of scarce resources, which could be employed elsewhere. As it is normally very costly or impossible to reverse such investment decisions, a rigorous and robust initial financial analysis is therefore, essential.

The value of a project is determined by the quantity and timing of the cash flows it generates over its lifespan. We saw in Chapter 3 how to estimate these cash flows as accurately as possible using techno-economic models. However, it is difficult to determine the value of a project or to compare the values of different projects by looking only at their cash flow projections. Financial analysis is the process of parameterising these cash flows into metrics which quantify project value. These can be compared to each other or to absolute threshold values, so that desirable projects can be identified. For this reason, the representativeness of projected cash flows is centrally important to financial appraisal and sound decision making: a financial parameter is only as good as the cash flows used in its calculation.

This chapter is divided into two main sections: fundamentals and financial measures. Fundamentals covers the main concepts required to calculate and understand financial measures. It stresses the importance of understanding the perspective of the investor in choosing relevant costs, benefits and the values of financial input parameters. The characteristics of investment decision types are outlined as these influence the choice of appropriate financial

Renewable Energy and Energy Efficiency: Assessment of Projects and Policies, First Edition.
Aidan Duffy, Martin Rogers and Lacour Ayompe.
© 2015 John Wiley & Sons, Ltd. Published 2015 by John Wiley & Sons, Ltd.
Companion Website: www.wiley.com/go/duffy/renewable

measures discussed in the second part of the chapter. A number of financial concepts that are important for adjusting cash flows to a form appropriate for financial appraisal are then introduced. These include the treatment of inflation, estimation of the present value of future cash flows with appropriate discount rates and the treatment of taxation and depreciation. Finally, project risk and uncertainty are addressed.

Several financial metrics are presented in the 'Financial Measures' section of the chapter. Some, such as the levelised cost of energy (LCOE) and savings-to-investment ratio (SIR), are specifically suited to energy project appraisal. Others, such as payback and discounted payback periods (PPs), return on investment (RoI), profitability index (PI), net present value (NPV), internal rate of return (IRR) and life cycle cost (LCC), are widely used across all industry sectors. The suitability of the measures for different project and decision types is presented at the end of the chapter.

4.2 Fundamentals

4.2.1 Investor perspective

We have already referred to the importance of 'investor perspective' – which is the viewpoint of the individual, group, company or society which the assessment is meant to be representative of – because of its fundamental importance to obtaining outputs which reflect the needs and wishes of that group. The same energy project will be viewed differently from different investor perspectives. For example, an electricity transmission system of overhead power lines will have different costs and benefits to:

- the network operator, which will wish to cover its capital costs through transmission use of system charges;
- the communities using the transmitted electricity, which will benefit due to domestic and industrial access to a secure energy supply; and
- the landowners and households along the transmission route, which will suffer from negative visual impacts, agricultural impacts and loss of property value.

As a different result will be obtained for each of these viewpoints, it is important to be clear from the outset what perspective is being represented. Although any project can be seen from many different perspectives, it is useful to distinguish between private and public viewpoints. A private perspective considers maximising the value to the firm for the benefit of its shareholders; therefore, only those additional costs and benefits directly attributable to the company resulting from the project are considered. Project costs and benefits that are 'external' to the company are ignored when considering a private perspective. These might include pollution costs or productivity benefits to wider society and are referred to as 'externalities' (because they are external to the assessment). A public perspective, however, concerns the impact of the overall welfare of the society being studied, usually expressed in monetary terms. As welfare is a broad measurement of societal well-being and includes many factors such as income, health, education and environment, a public perspective, therefore, attempts to consider all costs and benefits, including those which are external from a private investment perspective. An assessment of renewable energy support projects and policies which have significant social impacts must consider these social (public and private) costs and benefits in order to demonstrate whether they deliver welfare benefits. Where the ratio of benefits to costs is maximised, it is deemed to be economically 'efficient'. However, equity as well as efficiency must be considered for such projects. Equity is lost where some individuals are relatively worse off than others as a result of the project or policy; in such situations, the losers should be compensated. Therefore, a broader range of criteria must be considered for projects and policies which have significant societal impacts. For very significant projects multiple perspectives – both private and public – may need to be considered.

In this chapter, we consider the private investor perspective only. However, private firms must sometimes consider the wider social impacts of their projects where there are significant welfare impacts. For example, a project such as a large biomass power station necessitates the local production of biomass and results in additional airborne pollution in the locality. The resulting changes in farming practices and air quality will have significant welfare impacts. Therefore, any analysis must address the needs of the project's private investors, of the local community and of broader society. This broader societal perspective is addressed in Chapter 5.

4.2.2   Types of projects and decisions

It is important for the assessor to be aware of both the interdependencies of the projects being assessed and the decision to be made because these will influence the choice of the most appropriate financial parameter to be used. Projects can be classified as either independent or interdependent. Independent projects are those where the decision of whether to proceed with one does not affect the financial performances of the others. Any number of projects under consideration can be selected for investment where sufficient funds are available. *Interdependent* projects are those where the accept/reject decision affects the cash flows and financial performances of the other projects. Three types of interdependent projects are described as follows.

- *Mutually exclusive projects* are those where the acceptance of one project precludes the acceptance of the others. One reason for this exclusivity is where physical constraints exist. For example, only a gas- or a coal-fired power plant may be chosen for a particular site. Another is where the projects result in the same output. For example, a company may be considering investing in a gas boiler, oil boiler or biomass boiler, each individually capable of producing the necessary heat for the site.

- *Complimentary projects* are those where the acceptance of one increases net cash flows in another. For example, the construction of a large-scale electricity storage facility may reduce wind farm curtailment, thus increasing cash flows for wind investments from increased wind energy sales.
- *Prerequisite or contingent projects* are those where the acceptance of one project depends on the acceptance of another. For example, the construction of an offshore wind farm may be contingent on upgrading nearby port facilities.

Complimentary and prerequisite projects should be assessed together so that all costs and benefits associated with the projects are accounted for and fully informed decisions can be taken. Once the projects have been merged, then they can be assessed as independent or mutually exclusive projects.

The types of decisions normally considered during project appraisal include

- *Accept/Reject.* Where one or a number of independent projects are being considered and the decision of whether or not to proceed with each individual project is made. For example, an energy service company is considering whether to upgrade the pump motors in its portfolio of 15 commercial buildings. On 10 sites, the project has financial benefits, and these are accepted for implementation. The remaining five sites are rejected.
- *Choose from Mutually Exclusive Alternatives.* Where a project is being chosen from a number of competing alternatives, the investor should maximise the benefits of the single project which can be chosen. For example, where three different power stations are being considered for the same site, each with differing scale and output, then the project with the greatest value should be chosen.
- *Ranking.* Where the best sub-set is chosen from a set of independent projects. For example, due to capital constraints, a wind developer wishes to identify the portfolio of sites to develop from limited investment funds, which will maximise the value of the firm. The most valuable projects with a cumulative cost up to the investment threshold are identified.

4.2.3   Cash flows

We have already stressed that cash flows are the most important inputs used in the financial appraisal of energy projects. The cash flows that are considered in financial appraisals are the inflows and outflows which occur only if the project proceeds, known as 'incremental' cash flows. When identifying incremental cash flows, the following points should be considered.

- *Ignore Sunk Costs.* These are costs which have already been incurred or committed to; therefore they cannot be recovered. The investors are liable for these costs whether or not the project proceeds; therefore, they are not incremental costs and are irrelevant for the purposes of financial assessment of the project. For example, if a feasibility study has already been undertaken for a project, then this cost should not appear in the cash flow because it has been incurred. The investment decision should be forward-looking and consider avoidable future costs only.

- *Include Opportunity Costs.* This is the cost of the best alternative foregone. If you do not proceed with the project, then alternative costs arise in the form of another project or as a 'do nothing' alternative. For example, a software development company investing in photovoltaics (PV) panels for their office building may consider the achievable return of the foregone opportunity to invest in a new software product. There, the opportunity cost is the return which could have been made by investing in a software product instead of the energy project. However, if they are committed to investing in 'green' technologies (perhaps for reasons of corporate and social responsibility), then they should compare the project to realistic alternative technology investments foregone, such as wind turbines or solar water heaters.

- *Consider Substitution.* The output from a project may result in the loss of revenues elsewhere in the organisation due to the substitution of one energy form by another. For example, a gas company investing in a district heating system may lose gas sales to those customers who agree to purchase heat from the project.

- *Over and Above Costs.* Net costs should be used where appropriate. For example, if an architect is considering specifying a more-energy-efficient cladding system for an office building, then the capital cost net of the originally specified cladding system should be used when calculating the payback resulting from energy savings.

Cash flows are typically lumped into 1-year time periods. Flows may occur at the beginning, middle or end of the period, but in keeping with common practice an end-of-period practice is adopted in this book. Another convention is for up-front investment costs to occur in 'year 0', representing a theoretical 'overnight' build up of the energy investment. This assumption is acceptable for small, simple projects such as the installation of a gas boiler, but for large, complex projects this is a gross simplification which could result in erroneous results because investment cost may be incurred over years or even decades. Nonetheless, for simplicity, this assumption is adopted for most of the examples given in this book. However, the phasing of capital costs and revenue streams can have a significant impact on project viability and, therefore, must be considered in any detailed analysis of the financial performance of large, complex or phased projects.

Cash flows can be categorised into three broad categories based on the main activities involved in an energy-related project: *investment cash flows*, *cash inflows* and *cash outflows*. The former is often termed the 'capital cost' of the project while inflows and outflows are often referred to as 'revenues' and 'operating' or 'operation and maintenance' (O&M) costs, respectively.

Energy supply and efficiency project cash inflows typically result from either the sale of energy or avoided energy costs or savings in operating costs. Energy cash inflows are the products of the energy produced (or avoided) and the relevant tariff, summed for each energy type (e.g. heat, electricity and cooling). For example, cash inflows for an embedded combined heat and power (CHP) system would include the value of gas or electricity displaced (the product of quantity of electricity imports avoided and the electricity tariff plus the product of boiler gas avoided and the gas tariff) as well as the value of any electricity exported to the grid. The positive tax effects of depreciation are also included as cash inflows. The treatment of taxation in cash flows is discussed in more detail in Section 4.2.7.

Cash outflows are recurrent cash outgoings which, in energy-related projects are typically fuel and O&M costs. For renewable energy supply and energy efficiency projects, however, fuel costs are normally absent because wind-, ocean- and solar-energy are free. Projects that use biofuel, however, have energy input costs. The costs of labour, replacement materials, rent and insurances are also important cash outflows for all energy supply projects but are often negligible for passive energy investments such as envelope insulation.

In energy supply and efficiency projects, investment cash flows are usually substantial non-repetitive cash flows resulting from the initial purchase of plant and equipment. Typically, they occur at the start of a project when the main plant and equipment must be purchased, but substantial investments may be made later in the project as capacity is added or further project phases are implemented. Investments may occur over 1 year or several years depending on the construction programme and project phasing and must be incremental, that is, considering the investment net of avoided costs and the sales of displaced equipment. For example, a CHP system may avoid the need for stand-by generation and allow some existing boilers to be sold off. Capital subsidies and the residual end-of-life project value should be included as negative investment cash flows (i.e. positive cash flows) in the appropriate time periods. Many energy projects have residual values at the end of the investment lifespan due to land assets (which may have valuable planning permissions and access to transmission infrastructure) and the scrap value of plant and equipment.

The difference between cash inflows and outflows is the *net operating cash flow* (see Equation 4.1). The sum of cash inflows, outflows and investment cash flows is termed the *net cash flow* and is given by Equation 4.2.

$$F_{no,n} = F_{i,n} - F_{o,n} \tag{4.1}$$

$$F_{n,n} = F_{i,n} - F_{o,n} - F_{c,n} = F_{no,n} - F_{c,n} \tag{4.2}$$

where $F_{no,n}$ is the net operating cash flow in investment year n, $F_{i,n}$ and $F_{o,n}$ are the sums of cash inflows and outflows, respectively, in year n, $F_{n,n}$ is the net cash flow in investment year n and $F_{c,n}$ is the investment cash flow in year n.

The lifespan of a project determines the operational period over which it will continue to generate revenues and is therefore an important parameter in any investment appraisal. Lifespan is usually chosen on the basis of those of other, comparable, projects. Typical investment lifespans are given in Chapter 2, although these are increasing for many technologies due to ongoing technical and operational advances.

Project finance costs are taken into account using the appropriate cost of capital or an appropriate discount rate, both of which are discussed later in this chapter. The financing mechanism is, therefore, separate from the investment decision, which is not considered in the investment appraisal process.

Example 4.1   Classifying Cash Flows

A large office is considering investing in a CHP plant on its premises. Following an economic analysis, an optimum electrical output of 2 MW (i.e. 2 MW(e)) is chosen for the site. The system has an estimated capital cost of €2.3m and will be located on two car parking spaces in the basement of the office. A number of heating pipework modifications costing €40,000 have already been made in anticipation of the project proceeding. It is estimated that each year the system will reduce grid electricity imports by 7358 MWh and that 10,092 MWh of heat output will be used by the office complex. In addition, 3154 MWh of electricity will be exported onto the national grid during periods of low site electrical demand. The tariff for electricity imported is 8.2 c/kWh and that for exported electricity is at a government-subsidised 12 c/kWh. The gas tariff for the existing boilers (which have a measured efficiency of 80%) is 4 c/kWh. It is assumed that the current gas and electricity tariffs will not change with the introduction of the CHP system. The cost of O&M is 2 c/kWh(e) while insurances will cost €15,400 per annum. The cost of renting a car parking space is €1200 per annum. The main components of the CHP system have a projected lifespan of 15 years after which the residual value of the system is estimated at €250,000. Classify these cash flows over the project lifespan.

Incremental capital cash flows of €2.3m (outflow) and €250,000 (inflow) occur in years 0 and 15, respectively. The cost of pipework modifications is disregarded as this has been incurred whether or not the project proceeds and is, therefore, a sunk cost. The sources of incremental cash inflows associated with the CHP project are the avoided boiler gas purchase cost, the avoided grid electricity purchase cost and the sales of surplus electricity to the grid. These are the product of the relevant energy tariffs and quantities of energy avoided or sold. It should be noted that an appropriate avoided heat tariff (rather than an avoided gas tariff) must be applied to the useful thermal energy output from the CHP. The boiler efficiency of 80% and gas tariff of 4 c/kWh result in an avoided boiler heat tariff of 4.7 c/kWh. Incremental cash outflows include the value of gas consumed by the CHP and maintenance costs. Annual operating cash inflows and outflows are shown in Table 4.1. The additional incremental costs of the car parking spaces and insurances must be included.

Table 4.1 Calculation of operational cash flows for the CHP system.

Cash Flow Type	Description	Quantity	Tariff	Value (€000/annum)
Cash inflows	Electricity imports displaced	7,358 MWh/annum	0.082 €/kWh	603
	Electricity exported	3,154 MWh/annum	0.120 €/kWh	378
	Boiler heat displaced	10,092 MWh/annum	0.047 €/kWh	474
Total cash inflows				1456
Cash outflows	Gas used by CHP	26,280 MWh/annum	0.034 €/kWh	894
	O&M	10,512 MWh/annum	0.020 €/kWh	210
	Insurances	1 policy/annum	15,300 €/policy	15
	Car parking spaces lost	2 spaces/annum	1,200 €/space	2
Total cash outflows				1121
Net operating cash flows				335

Table 4.1 summarizes the classification and calculation of all operational cash flows in accordance with Equation 4.1. Cash flows over the lifespan of the CHP project are shown in Table 4.2.

Table 4.2 Summary cash flows over the lifespan of the CHP system (negative values in parenthesis).

	Year															
	0	1	2	3	4	5	6	7	8	9	10	11	12	13	14	15
Investment cash flows	(2300)															250
Total cash inflows		1456	1456	1456	1456	1456	1456	1456	1456	1456	1456	1456	1456	1456	1456	1456
Total cash outflows		(1121)	(1121)	(1121)	(1121)	(1121)	(1121)	(1121)	(1121)	(1121)	(1121)	(1121)	(1121)	(1121)	(1121)	(1121)
Net cash flow	(2300)	335	335	335	335	335	335	335	335	335	335	335	335	335	335	585

4.2.4 Real and nominal prices

Cash flows can be reported in either 'nominal' or 'real' prices. Nominal prices are those which we are used to dealing with in our day-to-day lives – they are the monetary values expressed in units of currency which increase (usually) from year to year due to inflationary effects. Real prices remove the effects of inflation from nominal prices by converting them to the prices of a 'base year', which is chosen to suit the needs of the study being conducted. Typically, the base year is the first year of the project but it may be the year the financial analysis is undertaken or any other suitable year. Real and nominal cash flows are related by

$$F_{r,j} = \frac{F_{n,i}}{(1 + e)^{(i-j)}} \tag{4.3}$$

where $F_{r,j}$ is the real cash flow equivalent for base year j, $F_{n,i}$ is the nominal cash flow in year i and e is the inflation rate.

Inflation is the percentage rate of change in price level over time which results in a decrease in purchasing power and is usually reported for annual periods. The central banks of many mature economies such as the United Kingdom, the Eurozone and the United States typically target inflation rates close to 2% when choosing economic policies, although actual inflation will vary around this target depending on factors such as fiscal and monetary policies as well as global and regional economic conditions. Inflation rates are normally presented in the form of price indices normalised to a base year, thus making conversions between nominal and real prices more convenient.

Example 4.2   Real and Nominal Prices

Calculate the real value in 2013 prices of €1.3m to be received in 2016 when inflation is expected to increase by 1.7% each year over the period. Substituting these values into Equation 4.3 we get

$$F_{r,2013} = \frac{1.3 \times 10^6}{(1 + 0.017)^{(2016-2013)}} = 1.236 \times 10^6$$

Therefore, the expected real value of €1.3m to be received in 2016 is €1.236m in 2013 prices due to the effects of projected inflation.

One such index is the consumer price index (CPI), which is widely used to measure price level changes over time in different regions. It is a measure of changes in price levels of a 'basket' of goods, which are representative of the economy-wide cost of living. Table 4.3 summarizes the EU's standardised CPI – the Harmonised Index of Consumer Prices (HICP) – for selected European countries and the United States normalised to a 2005 base year. CPI is used to convert nominal prices to real prices, as shown in Equation 4.4. Other more specific measures of inflation exist, for example, inflation indices for individual economic sectors include construction, producer (manufacturing) and services sector price indices. The choice of index depends on the cash flows being considered: a construction price index may be most appropriate for the construction of a dam where civil engineering costs dominate, whereas a manufacturing price index may be more appropriate for an onshore wind farm where turbine costs are most important. Real and nominal prices are related by a CPI according to

$$F_{r,i} = \frac{F_{n,i}}{CPI_i} \tag{4.4}$$

where $F_{r,i}$ is the real cash flow in year i converted to the base year, $F_{n,i}$ is the nominal cash flow in year i and CPI_i is the CPI in year i normalised to the base year.

Table 4.3   Harmonised Indices of Consumer Prices (HICP) for selected European countries and the United States (HICP 2005 = 100).

Location	2003	2004	2005	2006	2007	2008	2009	2010	2011	2012
EU-28	95.6	97.8	100.0	102.3	104.7	108.6	109.6	111.9	115.4	118.4
Bulgaria	88.8	94.3	100.0	107.4	115.6	129.4	132.6	136.6	141.2	144.6
Czech Republic	96.0	98.4	100.0	102.1	105.1	111.7	112.4	113.7	116.2	120.3
Denmark	97.5	98.3	100.0	101.8	103.5	107.3	108.4	110.8	113.8	116.5
Germany	96.4	98.1	100.0	101.8	104.1	107.0	107.2	108.4	111.1	113.5
Ireland	95.7	97.9	100.0	102.7	105.6	108.9	107.1	105.4	106.6	108.7
Greece	93.8	96.6	100.0	103.3	106.4	110.9	112.4	117.7	121.4	122.6
Spain	93.9	96.7	100.0	103.6	106.5	110.9	110.6	112.9	116.4	119.2
France	95.9	98.1	100.0	101.9	103.6	106.8	106.9	108.8	111.3	113.8
Italy	95.7	97.8	100.0	102.2	104.3	108.0	108.8	110.6	113.8	117.5
Netherlands	97.2	98.5	100.0	101.7	103.3	105.5	106.6	107.6	110.2	113.3
Austria	96.1	97.9	100.0	101.7	103.9	107.3	107.7	109.5	113.4	116.3
Poland	94.5	97.9	100.0	101.3	103.9	108.3	112.6	115.6	120.1	124.5
United Kingdom	96.7	98.0	100.0	102.3	104.7	108.5	110.8	114.5	119.6	123.0
Norway	97.9	98.5	100.0	102.5	103.2	106.7	109.2	111.8	113.1	113.6
United States	93.9	96.4	100.0	103.2	105.9	110.5	109.6	112.3	116.6	119.0

Source: Eurostat.

Example 4.3   Real and Nominal Prices

Calculate the real value in 2006 prices of €2.1 m spent in 2010 in the Netherlands. Using HICP presented in Table 4.3 (2005 base year), we see that the Dutch CPIs for 2006 and 2010 are 101.7 and 107.6, respectively. Therefore, normalising to a base year of 2006, the Dutch CPI for 2010 becomes 105.8((107.6/101.7) × 100). Substituting into Equation 4.4 we get

$$F_{r,2005} = \frac{2.1 \times 10^6}{105.8} = 1.985 \times 10^6$$

So the real 2006 cost of a nominal €2.1 m paid in 2010 is €1.985 m.

4.2.5   Present value

In general, people (and therefore the companies they own) prefer to receive cash today than in the future. For example, when given the choice, an individual would prefer to receive €100 today rather than in a year's time. In order to make it attractive to defer the receipt of this money, one might have to offer to give the individual €120 or more next year instead of €100 today. Therefore, the receipt of €100 in 1 year is less valuable to them as compared to its receipt today; it is worth up to €83.33 (€100 × 100/120) in today's terms. This is the present value of the future €100 cash flow, which is a function of the future cash flow and an annual discount rate.

In general, the present value of a future cash flow is given by

$$F_{pv,n} = \frac{F_{fv,n}}{(1+r)^n} \tag{4.5}$$

where $F_{pv,n}$ is the present value of the future cash flow $F_{fv,n}$ in n years' time and r is the discount rate.

Using this relationship, annual cash flows can be discounted to give present values for each year over the project's lifespan. This is typically executed using spread sheet software. However, in the case of manual calculations, rather than computing the denominator in Equation 4.5 for each present value calculation, a *discount factor* (DF) can be used instead. A discount factor is the factor by which a future cash flow must be multiplied to give its present value. It is a function of the discount rate and the time period in which the cash flow occurs. A table of discount factors can be found at the end of this book in Appendix A. It is given by

$$DF_{n,r} = \frac{1}{(1+r)^n} \tag{4.6}$$

so that

$$F_{pv,n} = F_{fv,n} \times DF_{n,r} \tag{4.7}$$

where DF_n is the discount factor in year n for the discount rate r and $F_{pv,n}$ is the present value of the future cash flow F_n in n years' time.

Example 4.4   Present Values of Future Cash Flows

What is the present value of cash flows of €20, €30 and €40 received respectively in 1, 2 and 3 years' time using a discount rate of 10%? We first calculate the present value of €20 received in 1 year's time with a 10% (0.1) discount rate using Equation 4.5:

$$F_{pv,1} = \frac{20}{(1+0.1)^1} = €18.18$$

Similarly, for years 2 and 3

$$F_{pv,2} = \frac{30}{(1+0.1)^2} = €24.79$$

$$F_{pv,3} = \frac{40}{(1+0.1)^3} = €30.05$$

Example 4.5   Discount Factors

A project has a lifespan of 5 years and is being evaluated using three possible discount rates: 5%, 10% and 15%. Calculate the discount factors which should be used to discount cash flows for each year and discount rate and use these to determine the present values of a recurrent €5m cash flow in each year.

We first use Equation 4.6 to calculate discount factors. For example, the discount factors for a discount rate of 5% in years 1–5 are given by

$$\text{Year 1: } DF_{1,0.05} = \frac{1}{(1+0.05)^1} = 0.95$$

$$\text{Year 2: } DF_{2,0.05} = \frac{1}{(1+0.05)^2} = 0.91$$

$$\text{Year 3: } DF_{3,0.05} = \frac{1}{(1+0.05)^3} = 0.86$$

$$\text{Year 4: } DF_{4,0.05} = \frac{1}{(1+0.05)^4} = 0.82$$

$$\text{Year 5: } DF_{5,0.05} = \frac{1}{(1+0.05)^5} = 0.78$$

Discount factors for the 10% and 15% discount rates are similarly calculated. Using Equation 4.7, the recurrent cash flow of €5m is then multiplied by each discount factor to obtain its present value for the relevant year. Results are shown in Table 4.4

Table 4.4   Discount factors and present values for a €5m (real) annual cash flow for 5 years.

	Discount rate	Year				
		1	2	3	4	5
Discount factor	5%	0.95	0.91	0.86	0.82	0.78
Present value (€m)		4.76	4.54	4.32	4.11	3.92
Discount factor	10%	0.91	0.83	0.75	0.68	0.62
Present value (€m)		4.55	4.13	3.76	3.42	3.10
Discount factor	15%	0.87	0.76	0.66	0.57	0.50
Present value (€m)		4.35	3.78	3.29	2.86	2.49

It is important to note that present value and inflation should not be confused. Inflation relates to the decline in the purchasing power of a sum of money due to increasing prices levels in an economy. In contrast, the present value of a real future cash flow is lower than that received today because the opportunity to invest it in another project has been lost. The difference between present and future cash flow values represents the cost of giving up an alternative investment over the time period being considered and is referred to as the time value of money. Future cash flows may be real or nominal and each must be discounted to today's worth using real and nominal discount factors; this is dealt with in the next section.

4.2.6   Discount rates

Future cash flows can be converted to present values using an appropriate discount rate. This rate is a measure of time value of money – the amount of return that could have been made in the next best investment opportunity foregone. The discount rate comprises both a 'risk-free' rate and a 'risk premium'. The former is what could be earned in a risk-free investment such as a high-quality government bond; the latter is the additional rate of earning required to compensate for the risk profile of the particular project in question. For example, an investor might require a higher return for an ocean energy investment than for a gas-fired power station given that the former is a less proven technology and therefore is a more risky investment. However, if the investor invested their money in a government bond they might accept a very low return, which would be close to the risk-free rate. The additional return required over this risk-free rate is the risk premium.

Discount rates may be nominal or real which include or exclude, respectively, the effects of inflation. Real discount rates are used in Equation 4.5 when real cash flows have been estimated; similarly, nominal discount rates are used with nominal cash flows. It is extremely important that the analyst is aware whether real or nominal cash flows are being used, and that the appropriate discount rate or factor is applied consistently. Never mix real discount rates and nominal cash flows and vice versa. Real and nominal rates are related by

$$(1 + d_n) = (1 + d_r)(1 + e) \qquad (4.8)$$

where d_n and d_r are the nominal and real discount rates, respectively, and e is the inflation rate.

Example 4.6   Discounting Cash Flows

A new commercial gas boiler costs €50,000 in year 0 with real annual net operating cash flows of €50,000 over 15 years when savings in fuel costs over the old oil boiler are taken into account. The company uses a nominal discount rate of 12% to assess its investments and inflation is projected to be 2% per annum over the investment period. Calculate the cash flows discounted to year 0. Ignore taxation and residual values.

Here the nominal discount rate of 12% cannot be applied to the real cash flows. Two approaches may be taken: first, a real discount rate can be calculated using Equation 4.8; second, the real cash flows are converted to nominal cash flows using the inflation rate and then discounted using the nominal discount rate. Both approaches are illustrated as follows.

Method 1. Substituting values into Equation 4.8 and rearranging we calculate the real discount rate as follows:

$$(1 + 0.12) = (1 + d_r)(1 + 0.02)$$

$$d_r = \left[\frac{(1 + 0.12)}{(1 + 0.02)}\right] - 1 = 0.098$$

The real discount rate is therefore 9.8% and this is used to determine the present values of the cash flows using Equation 4.5. Results are shown in Table 4.5.

Method 2. Real cash flows are inflated using Equation 4.3. Equation 4.5 is used to discount the resulting nominal cash flows to present values using the nominal discount rate. Results are shown in Table 4.5.

Table 4.5   Discounted cash flows calculated using two different methods as outlined in Example 4.6.

	Year					
	0	1	2	3	4	5
Real undiscounted cash flow (€)	50,000	15,000	15,000	15,000	15,000	15,000
Method 1						
Discounted cash flow using real discount rate (9.8%)	50,000	13,661	12,441	11,330	10,319	9,397
Method 2						
Nominal undiscounted cash flow (2% inflation)	50,000	15,300	15,606	15,918	16,236	16,561
Discounted cash flow using nominal discount rate (12%)	50,000	13,661	12,441	11,330	10,319	9,397

The question arises as to what discount rate to apply when assessing the economic performance of an energy project. The answer is: it depends, because discount rates vary between industry sectors, companies, projects and individuals. For example, a study in Canada found minimum acceptable rates of return for a range of companies to be in the region of 5–40% depending on the industry sector, company activity and size. Companies, particularly large utilities, may use a discount rate called the weighted average cost of capital (WACC), which is equivalent to the minimum return which must be made to pay all its funders such as shareholders, banks and bondholders (depending on the funding structure of the company). This is the opportunity cost of capital to the company and should be used where the risk profile of the project is similar to that of the portfolio of projects owned by the company. Sometimes, the company will set the WACC as its 'hurdle rate' or 'minimum acceptable rate of return' (MARR), which must be at least equalled by the proposed project in order for it to be funded by the company. WACC is widely used by energy utilities for investment appraisal.

There are a number of methods for calculating WACC, but for a company with a simple funding structure comprising debt (e.g. in the form of bank loans) and equity (as shareholding) only and ignoring tax effects, it is given by the weighted average of the return on equity and cost of debt:

$$\text{WACC} = r_e \left[\frac{E}{(D+E)} \right] + r_d \left[\frac{D}{(D+E)} \right] \tag{4.9}$$

where r_e and r_d are the returns on equity and debt interest rates, respectively; E is the amount of equity (€) and D is the amount of debt (€).

In the case where debt servicing is tax-deductible and the company is funded with a mixture of common and preferred shares, Equation 4.9 can be written as

$$\text{WACC} = r_{eo} \left[\frac{E_o}{(D+E_o+E_p)} \right] + r_{ep} \left[\frac{E_p}{(D+E_o+E_p)} \right]$$
$$+ (1-T)r_d \left[\frac{D}{(D+E_o+E_p)} \right] \tag{4.10}$$

where r_{eo} and r_{ep} are returns on ordinary and preferred shares, respectively; E_o and E_p are the amounts of ordinary and preferred equity (€) and T is the corporate tax rate.

Example 4.7   Weighted Average Cost of Capital

A company is funded with €220m of debt and €130m of equity. If the real cost of debt is 6% per annum and the return on equity is 12%, calculate its WACC.

Using Equation 4.9

$$\text{WACC} = 0.12 \times \left[\frac{130}{(220+130)} \right] + 0.06 \times \left[\frac{220}{(220+130)} \right] = 8.23\%$$

Calculating the WACC for a company can be a complex task drawing on data from a variety of sources including company annual returns (for data such as debt and shares issued), banks (risk free rate) and financial data providers (equity and market risk). Where a project is a once-off investment or where no data are available for estimating WACC, the discount rate should be set at the opportunity cost of capital, which is the expected return on the next best alternative use of the available funds. The choice of this alternative depends on the investor and the risk profiles of the alternatives which should be similar for a fair comparison. For a company which specialises in installing wind farms, for example, it would be another wind farm project. However, a manufacturing company may have to choose between installing a new energy efficient boiler and a non-energy project such as upgrading a production line. For some companies, however, investing in energy efficient projects forms part of their environmental strategy and, as such, lower MARRs than those used to assess core activities may be applicable.

The discount rate for an individual must be considered when assessing the economic attractiveness of energy projects to consumers investing in a personal capacity on projects such as home energy improvements or the purchase of electrical vehicles. In this instance, it is difficult to choose an appropriate value because individuals may have very different discount rates depending on factors such as age, gender and socio-economic status. Various studies (inter alia Hausman, 1979; Warner and Pleeter, 2001) have estimated individual personal discount rates of between 0% and 89% for decisions such as the purchase of energy-saving air conditioners or whether to defer lump sum severance payments. A typical value of 20% is suggested for energy-related investments.

One feature of discounting with a fixed discount rate is that as cash flows become more distant, their present value rapidly diminishes in an exponential fashion. For example, an annual net cash flow of €1m in 30 years' time is worth €231,000 using a 5% discount rate. Higher discount rates result in a more rapid erosion of present value. The same cash flow would be worth just €15,000 in 30 years using a 15% discount rate. The present values of a €1m annual net cash flow is shown in Figure 4.1 for discount rates of 5%, 10% and 15% applied over 30 years.

4.2.7   Taxation and depreciation

Depreciation is an important consideration in the financial assessment of capital-intensive investments such as many energy projects because – depending on its tax treatment – it can have a significant impact on cash flows. Depreciation is the decrease in the value of the asset over its useful life. A wind turbine, for example, will be subject to wear and tear over its productive lifespan, and its value will diminish accordingly until it can be considered worthless at the end of its operational life of 20 years or so. The treatment of depreciation in accounting does not always truly reflect this reduction in asset value. Specific rules govern depreciation in accounting where the value of assets may be reduced annually by prescribed amounts which may not match technology lifespan. These rules vary from country to country and should be determined as part of any economic appraisal project and be reflected in after-tax cash flows. In the United Kingdom, for example, wind power assets can be depreciated over an accelerated 5-year period whereas

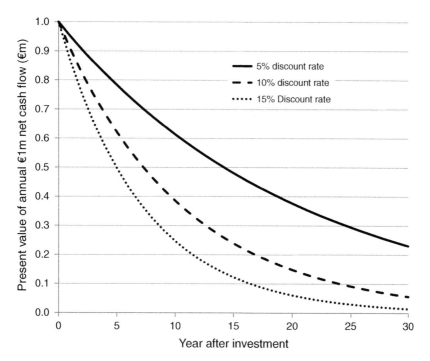

Figure 4.1   The present values of a €1m annual net cash flow over 30 years for discount rates of 5%, 10% and 15%.

this is not the case in the United States. However, the depreciation of energy supply and efficient technologies is only important to financial appraisal in its effects on project cash flows. In many countries annual depreciation can be written off against corporate tax liabilities which, subject to profitability, results in increased project cash flows in the relevant years.

There are a variety of depreciation methods that can be divided into two broad categories: accelerated and straight-line depreciation. Straight line is where the book value of the asset decreases to zero in equal amounts over the appropriate accounting period. Accelerated depreciation involves writing off more of the asset in the early years than later on, thus increasing the value of the project since the resulting increase in net cash flows earlier in the project have higher present values than without the accelerated tax benefits. Apart from the method used, two parameters are important: the asset's depreciable value and the depreciable life. The lifespan for depreciation purposes is not the lifespan used for estimating cash flows – the former is set by accounting rules, the latter based on expected lifespan.

Corporation tax (also known as 'company tax') is generally levied on company profits at a rate which typically ranges from 10% to 30% depending on the jurisdiction. The relevant tax rate should be taken into account when calculating after tax cash flows. However, depreciation can be deducted from company profits thus reducing corporate tax liabilities. Advice on national rules governing depreciation and other relevant taxes should be obtained from a tax specialist or accountant as part of any detailed energy project appraisal.

Example 4.8   Tax Effects of Depreciation

A cement manufacturer plans to upgrade their boiler plant to make it more energy efficient. The cost of the investment is estimated to be €800,000, and it will qualify for accelerated straight-line depreciation over 5 years, although the investment has a projected operational lifespan of 12 years. Net real operating cash flows are €60,000 for each year of operation. If the company pays a marginal corporation tax rate of 20% on its profits, calculate the after-tax cash flows for the project assuming the company generates sufficient profits to write off all depreciation.

Straight-line depreciation of the asset results in a depreciation expense of €160,000 per annum. Setting this against the €60,000 annual profit gives a net loss of €100,000 per annum for the first 5 years. This provides a €20,000 tax credit for the company (at a 20% tax rate), and assuming the company's pre-tax profits are greater than this amount, it can be fully deducted giving annual incremental project after-tax cash flows of €80,000 for the first 5 years. Thereafter no further depreciation occurs and project profits are taxed at 20% to give an after-tax net cash flow of €48,000. This is illustrated in Table 4.6 where the €20,000 tax credit is shown in brackets as a negative tax.

Table 4.6   Effect of depreciation and corporation tax on after-tax net cash flows for Example 4.8 (all figures in €000s).

						Year							
	0	1	2	3	4	5	6	7	8	9	10	11	12
Before tax cash flows	(800)	60	60	60	60	60	60	60	60	60	60	60	60
Depreciation		(160)	(160)	(160)	(160)	(160)							
Taxable income		(100)	(100)	(100)	(100)	(100)	60	60	60	60	60	60	60
Tax (20%)		20	20	20	20	20	(12)	(12)	(12)	(12)	(12)	(12)	(12)
After tax cash flows	(800)	80	80	80	80	80	48	48	48	48	48	48	48

4.2.8   Unequal project lifespan

Investment appraisal of mutually exclusive projects may involve comparing projects which have different lifespans. This is problematic because the same length of service is not considered in both cases and the projects cannot, therefore, be directly compared to one another. There are two main approaches to this problem: choose the higher equivalent annual cost (EAC) or calculate the financial parameter using the least common multiple method.

EAC is the present value of the annual cost of investing in and operating the energy project over its lifespan. It is given by

$$\text{EAC} = \frac{\text{NPV}}{A_{N,d}} \tag{4.11}$$

where NPV is the net present value of the project (see Section 4.3.4), A is the annuity factor, N is the project lifespan and d is the discount rate. A is given by

$$A_{N,d} = \frac{1 - (1 + d)^{-N}}{d} \tag{4.12}$$

Example 4.9   Equivalent Annual Cost for two projects with unequal lives

Two wave energy technologies are being assessed for a 10 MW wave energy farm. Project A requires an initial capital investment of €50m with net operating cash flows of €6m over a 15-year lifespan. Project B involves €40m with net operating cash flows of €7m over a 10-year lifespan. Use the EAC to establish which project is more attractive to the investor. Assume a real discount rate of 8% and disregard tax and residual project values.

Using Equation 4.12, we calculate the annuity factors for both projects as

$$A_A = \frac{1 - (1 + 0.08)^{-15}}{0.08} = 8.56$$

$$A_B = \frac{1 - (1 + 0.08)^{-10}}{0.08} = 6.71$$

Using Equation 4.19, we calculate the NPV for both projects:

$$\text{NPV}_A = -50 + \sum_{n=1}^{15} \frac{6}{(1 + 0.08)^n} = €1.36m$$

$$\text{NPV}_B = -40 + \sum_{n=1}^{10} \frac{7}{(1 + 0.08)^n} = €6.97m$$

EACs are given by Equation 4.11:

$$\text{EAC}_A = \frac{1.36}{8.56} = €0.16m$$

$$\text{EAC}_B = \frac{6.97}{6.71} = €1.04m$$

Using this measure, Project B is a more attractive investment because it has a higher EAC.

The least common multiple method (sometimes referred to as the replacement chain method) involves identifying the least common multiple of the project lifespan being considered and repeating life cycle project cash flows up to this multiple.

Example 4.10   Least Common Multiple approach to comparing unequal project lives

Use the least common multiple approach to create two comparable cash flows as in Example 4.9 for further financial analysis.

The least common multiple of 10 and 15 is 30 years, so Project A will be repeated twice and Project B three times. Investments for Project A will occur in years 0 and 15 with net operating cash flows of €6m in years 1–30. In year 15, the net cash outflow is €44m which is the difference between the net operating cash flow (€6m) and the investment (€50m) in that year. Investments for Project B will occur in years 0, 10 and 20 with net operating cash flows of €7m in years 1–30. In years 10 and 20 the net cash flow is the difference between the net operating cash flows (€7m) and the investments (€40m) in those years (Table 4.7).

Table 4.7   Least common multiple approach to creating comparable cash flows for Example 4.10.

	Year																														
	0	1	2	3	4	5	6	7	8	9	10	11	12	13	14	15	16	17	18	19	20	21	22	23	24	25	26	27	28	29	30
Project A	(50)	6	6	6	6	6	6	6	6	6	6	6	6	6	6	(44)	6	6	6	6	6	6	6	6	6	6	6	6	6	6	6
Project B	(40)	7	7	7	7	7	7	7	7	7	(33)	7	7	7	7	7	7	7	7	7	(33)	7	7	7	7	7	7	7	7	7	7

4.3   Financial measures

When assessing investments, the management need a way to identify acceptable projects, reject unacceptable ones or postpone projects, which may be attractive at some future date. In addition, they need to rank acceptable projects in order of value to the shareholder. There are a wide variety of financial measures which can be used to value and compare energy projects. They include

- *financial ratios* such as the SIR, RoI and PI;
- *temporal measures* such as the simple PP and the discounted PP;
- *rates of return* such as the IRR and the modified internal rate of return (MIRR);
- *cost metrics* such as LCCs and the LCOE; and
- *aggregated cash values* such as the NPV.

Some of these measures use the real, undiscounted cash flow values and, as we saw previously, such approaches are inferior to those using discounted cash flows. Although different measures have their particular strengths and weaknesses, NPV is widely regarded as the best measure for maximising shareholder value, and it is recommended that this measure be applied whenever possible. Rates of return are popular because they can be directly compared to other investment products or to threshold returns required by company management. Financial ratios indicate the ratio of various benefits to costs. Cost metrics are useful for comparing projects with identical outputs where the least-cost option is being sought. Temporal measures can identify how long an investment is at risk before payback.

In this section, the methods for calculating these financial metrics are described and their strengths and weaknesses are discussed. They should be used to aid rather than be the only basis for decision making as energy projects are often complex and involve more than just financial considerations. Informed decisions can only be made by management if they are fully aware of the assumptions made, so these should be clearly communicated as part of the reporting process. A risk analysis should form part of the analysis for all but the simplest projects. Normally, more than one financial parameter is used in the decision-making process. In practice, whichever financial parameters are adopted must be applied consistently to all alternatives being reviewed.

4.3.1   Payback and discounted payback periods

The PP (or simple PP) is the time required to recover a project investment, usually expressed in years. It is probably the most well-known and popular financial measure because it is so easy to calculate and understand. A project 'pays for itself' at a point in time when the cumulative net cash flows equal the net investment cash flows; the time from the beginning of the project to reach this point is termed the 'payback period'. For projects which have equal annual net operating cash inflows (where annual cash outflows and inflows remain steady), the PP is given by

$$n_p = \frac{F_c}{F_{no}}, \quad \text{for } F_{no} > 0 \tag{4.13}$$

where n_p is the simple PP (years), F_c is the initial capital investment and F_{no} is the real undiscounted annual net operating cash flow.

Example 4.11   Payback period with equal annual net operating cash flows

A proposed 10 MW PV Farm is projected to generate 15 GWh/annum and cost €20 million to build. The investors can avail of a long-term government feed-in-tariff of €150/MWh for all output from the site. If the annual operating costs are €100,000, calculate the projected PP for the project.

The real operational cash inflows are the product of the annual output and feed-in-tariff giving €2.25m/annum. Real annual operating outflows of

€100,000 result in a net operating cash flow of €2.15m/annum. Substituting this and the capital cost of €20m into Equation 4.13, the project PP is

$$n_p = \frac{20}{2.15} = 9.3 \text{ years}$$

However, most projects do not have equal annual net operating cash flows. In these cases, the PP is the time period, n, after the project commencement when the sum of the net investment cash flows equals the net cash inflows:

$$\sum_n F_{c,n} = \sum_n F_{no,n} \qquad (4.14)$$

rearranging

$$\sum_n F_{c,n} - \sum_n F_{no,n} = 0 \qquad (4.15)$$

For the PP, $F_{c,n}$ is the investment cash flow in time period n and $F_{no,n}$ is the net cash flow in period n.

Example 4.12 Payback Period with Unequal Net Annual Cash Flows

A company is assessing whether to install a biomass-fired boiler instead of a conventional oil-fired boiler to heat its premises. The biomass boiler has an over-and-above (incremental) capital cost of €150,000 compared to the oil boiler. The company is new and growing and plans to occupy the building on a phased basis over 7 years; as a result, annual incremental operating savings (net operating cash flows) begin at €10,000 in year 1 and rise to €60,000 in year 7 as shown in Table 4.8. The system lifespan is projected to be 15 years. What is the PP?

Table 4.8 Incremental cash flows for Example 4.12 showing a PP in year 6 for a project with unequal net annual cash flows (all figures in €000s).

	Year															
	0	1	2	3	4	5	6	7	8	9	10	11	12	13	14	15
Incremental capital cost of biomass boiler system	(150)						Project pays back in year 6									
Annual incremental system savings		10	22	30	37	40	51	60	60	60	60	60	60	60	60	60
Cumulative cash flows	(150)	(140)	(118)	(88)	(51)	(11)	**40**	100	160	220	280	340	400	460	520	580

Applying Equation 4.15 the period in which the incremental investment cash flows equal the incremental net operating cash flows is identified. Table 4.8 shows that between the end of years 5 and 6 the cumulative net cash flow switches from negative to positive, indicating that incremental investment and net operational cash flows equal one another at some point during this period. Interpolating linearly gives a PP of 5.22 years.

A significant drawback of the PP method is that it does not take the time value of money into account. While this does not have a significant impact on the accuracy of short PPs with low discount rates, it results in misleading values for the longer PPs which are typical of many energy-related projects. The discounted payback period (DPP) overcomes this shortcoming by adjusting cash flows to account for the time value of money. This normally has the effect of increasing the time to payback so that the DPP is longer than the PP. Given that discounted net operating cash flows will vary from year to-year, Equation 4.14 is used to calculate the DPP using discounted cash flows where the DPP $F_{c,n}$ and $F_{no,n}$ are discounted cash flows.

Example 4.13   Discounted Payback Period

The management of the company referred to in Example 4.12 does not wish to make a decision based on a PP which does not consider the time value of money. What is the DPP for the biomass project using a real discount rate of 8%?

The annual net cash flows are discounted using the 8% discount rate to give the discounted net cash flows as shown in Table 4.9. The point in time at which these sum to zero is the DPP, which is found to be during year 7. Interpolating gives a DPP of 6.33 years.

Table 4.9   Discounted cash flows for Example 4.13 showing the discounted payback occurring in year 7.

							Year								
	0	1	2	3	4	5	6	7	8	9	10	11	12	13	14	15
Incremental capital cost of biomass boiler system	(150)								Project pays back in year 7							
Annual incremental system Savings		10	22	30	37	40	51	60	60	60	60	60	60	60	60	60
Net cash flows	(150)	10	22	30	37	40	51	60	60	60	60	60	60	60	60	60
Discounted net cash flows (8%)	(150)	9	19	24	27	27	32	35	32	30	28	26	24	22	20	19
Cumulative discounted net cash flows	(150)	(141)	(122)	(98)	(71)	(44)	(12)	24	56	86	114	139	163	185	206	225

Both PP and DPP suffer from significant weaknesses as measures of the attractiveness of an investment. In reality, neither method provides a measure of profitability or investment attractiveness because it ignores the value of all project cash flows after the PP is reached. They are really measures of liquidity – how quickly the asset can be converted to cash by repaying the initial investment. In this regard, they can be used as a measure of how long an investment is at risk before repayment. Neither the undiscounted nor discounted method takes the project size into account; therefore, the economic performance – as measured by PP – of a very small project can be indistinguishable from that of a very large one although they offer very different investment opportunities. Therefore, although PP and Discounted PP are commonly used and easy to calculate, they can be a poor measure of the value of a project and should be used only as a secondary measure to a metric such as NPV, which is discussed later in this chapter.

4.3.2 Return on investment

The RoI is the ratio of annual after-tax income to the project investment. There are a number of different approaches, but the simplest involves dividing average net annual operating cash flow by the total investment cash flows:

$$\text{RoI} = \left(\frac{F_{no}}{F_i} \right) \times 100 \qquad (4.16)$$

where RoI is the return on investment, F_{no} is the average recurrent undiscounted real after-tax net operating cash flow (€) and F_i is the initial capital investment net of any capital subsidies.

Example 4.14   Return on Investment

An offshore wind farm comprising 22 turbines each rated at 3.6 MW is projected to cost €333.4m, with energy sales of €32m/annum and operating expenses of €7m/annum. The initial investment is subject to straight-line depreciation over 15 years which can be written off against taxable income at a corporation tax rate of 18%. What is the projected return on investment for the project?

Table 4.10 shows annual cash inflows (from energy sales), cash outflows (cost of sales) and investment cash flows (capital cost). Inflows less outflows give net operating cash flows before tax, from which depreciation (capital cost spread evenly over 15 years) is deducted to give the annual taxable income from the project. This is used to calculate the corporation tax liability (by multiplying by 18%) which is deducted from net operating cash flows before tax to give the real undiscounted after tax net operating cash flow. The average of these flows over the 20-year investment period is €24m/annum. Using this figure together with the initial capital investment, Equation 4.16 gives

$$\text{RoI} = \left(\frac{24}{333} \right) \times 100 = 7.2\%$$

Table 4.10   Before and after tax annual cash flows for Example 4.14 (all figures in €m).

											Year										
	0	1	2	3	4	5	6	7	8	9	10	11	12	13	14	15	16	17	18	19	20
Capital cost	(333)																				
Energy sales		32	32	32	32	32	32	32	32	32	32	32	32	32	32	32	32	32	32	32	32
Cost of sales		(7)	(7)	(7)	(7)	(7)	(7)	(7)	(7)	(7)	(7)	(7)	(7)	(7)	(7)	(7)	(7)	(7)	(7)	(7)	(7)
Net operating profit (loss) before tax		26	26	26	26	26	26	26	26	26	26	26	26	26	26	26	26	26	26	26	26
Depreciation		(22)	(22)	(22)	(22)	(22)	(22)	(22)	(22)	(22)	(22)	(22)	(22)	(22)	(22)	(22)	0	0	0	0	0
Taxable income		3	3	3	3	3	3	3	3	3	3	3	3	3	3	3	26	26	26	26	26
Corporation tax (18%)		(1)	(1)	(1)	(1)	(1)	(1)	(1)	(1)	(1)	(1)	(1)	(1)	(1)	(1)	(1)	(5)	(5)	(5)	(5)	(5)
Net operating profit after tax		25	25	25	25	25	25	25	25	25	25	25	25	25	25	25	21	21	21	21	21

RoI is a simple measure and is useful for comparing to similar ratios for other investment opportunities and/or to the return on capital employed (ROCE) by the company, where relevant. A disadvantage of RoI is that it is not useful as an absolute measure of financial performance and cannot be used in isolation. There is no absolute threshold RoI, which indicates a project is an attractive investment (unlike, for example, NPV or PI). A second drawback of the approach is that it does not consider the time value of money so a project with 'front-loaded' net cash flows can give the same result as one where the majority of the cash flows occur at the end of the project; the former, however, is more valuable to investors since they receive cash more quickly. Finally, RoI only considers the accounting book value of the asset, rather than its market value. This has the effect of giving better results for assets which have higher market than book values and vice versa for those with lower market values.

4.3.3   Profitability index and savings-to-investment ratio

PI is the ratio of the discounted projected cash inflows to outflows. It is also referred to as cost–benefit ratio or benefit–cost ratio.[1] PI is given by

$$PI = \frac{PV(F_{no,n})}{PV(F_{i,n})} \tag{4.17}$$

$$PI = \frac{\sum_{n=0}^{N} F_{no,n}/(1+d)^n}{\sum_{n=0}^{N} F_{i,n}/(1+d)^n} \tag{4.18}$$

---

[1] In Chapter 6 we use the term cost–benefit analysis and cost–benefit ratio in assessing the value of projects or policies to society, rather than to the firm as we do in this chapter.

where $F_{no,n}$ is the net operating cash flow in year n, $F_{i,n}$ is the investment cash flow in year n, N is the analysis period and d is the annual discount rate.

A PI greater than unity indicates a desirable project as the present value of the net benefits is greater than the costs. The greater the value of PI above 1 the more attractive the project.

Example 4.15   Profitability Index

A 150 MW pump-hydro storage (PHS) system is budgeted to cost €165m to design and build and is expected to have an operational life of 40 years. It is estimated to produce and consume 329,000 MWh and 386,000 MWh of electricity per annum. Estimated operational cash flows are presented in Table 4.11 based on unit electricity costs of 20€/MWh, maintenance costs of 2€/MWh and electricity sales of 50€/MWh.

Table 4.11   Estimated operational cash flows for Example 4.15.

Cash flow	Years	Value (€m/annum)
Investment cost	0	165
Energy costs	1–40	7.73
Maintenance and Overhead costs	1–40	0.66
Revenues	1–40	16.43
Net operating cash flow	1–40	8.04

The discount rate to be applied to the project is 4% (real) over its lifespan. Substituting this and the values from Table 4.11 into Equation 4.18 we get

$$PI = \frac{\displaystyle\sum_{n=0}^{40} 8.04/(1+0.04)^n}{165} = \frac{159.1}{165.0} = 0.964$$

The PI is less than unity so is not an attractive investment in its current form.

Where net operating cash flows are the result of energy savings rather than revenues, the term savings-to-investment ratio (SIR) is often used instead of PI. The SIR is the ratio of the present value of net savings (equivalent to net operating cash flows) to the present value of incremental investments. As with PI, the higher the SIR, the greater the net saving per unit investment. For example, upgrading insulation results in reduced energy bills and an embedded CHP system, which supplies heat and electricity to the site where it is located will displace boiler fuel and grid electricity, thus reducing costs. Neither results in increased energy sales and both are therefore suitable for analysis using SIR.

SIR is given by Equations 4.17 and 4.18 where $F_{no,n}$ is the net operating cash flow *saving* in year n. Net savings typically involve displaced energy and O&M

costs. For example, a local government which replaces high-pressure sodium (HPS) street lighting with light-emitting diodes (LED) will not only benefit from lower energy costs but also from reduced maintenance and replacement costs due to the longer lifespan of the new technology. These savings are set against incremental investment costs, which include project capital costs, plus replacement costs, less avoided capital costs and residual value.

Example 4.16   Savings-to-Investment Ratio

An industrial process uses water flowing through a pipe at $17 \, m^3/min$, which is pumped using an 80-kW pump. It is proposed to replace the throttle valve that currently regulates flow by fitting the pump with a more efficient variable speed drive (VSD). The projected installed and commissioned cost of the new VSD is €18,200, and it is expected to result in savings of 210 MWh/annum at a cost of 9 c/kWh The company's hurdle rate is 9% and the VSD is expected to last 15 years. What is the SIR for the project?

Taking total annual savings of €18,922 (210 MWh × 8760h × 9 c/kWh) and a cumulative discount factor (see Section 4.3.4) of 8.061 from the discount factor table in Appendix A (9% discount rate and 15-year lifespan), the present value of all net savings is €152,527 (€18,922 × 8.061). The present value of capital investment is €18,200 given that it occurs at the very beginning of the project. The SIR is given by

$$\text{SIR} = \frac{152,527}{18,200} = 8.4$$

As this is significantly greater than unity, the project should be approved based on this measure.

As both PI and SIR are ratios they do not distinguish between different investment sizes and, therefore, should not be used for choosing between mutually exclusive projects, although they can be used to rank the relative attractiveness of projects, SIR being used where cost savings dominate.

4.3.4   Net present value

NPV measures the economic value of an investment as the sum of all discounted future net cash flows. All future cash inflows, cash outflows and investment cash flows are discounted to the base year using an appropriate discount rate and summed (see Equation 4.5). Where the NPV is zero or positive, then the project will give a return equal to or greater than the required discount rate and therefore the project is attractive to the investor; the greater the NPV, the more attractive the project. If it is negative, then the project should not proceed because it does not provide the minimum return required. However, in some situations projects with negative NPVs are accepted. These include

- mandatory projects such as those required by regulation;
- projects providing necessary outputs which are difficult to value; and
- projects which meet a company's policy objectives such as 'green' projects forming part of a company's environmental strategy.

NPV is related to net cash flows and discount rate by the equation

$$\text{NPV} = \sum_{n=0}^{N} \frac{F_{n,n}}{(1+d)^n} = F_0 + \frac{F_1}{(1+d)^1} + \frac{F_2}{(1+d)^2} + \cdots + \frac{F_N}{(1+d)^N} \quad (4.19)$$

where $F_{n,n}$ is the net cash flow in year n (see Equation 4.1), N is the analysis period (typically the project lifespan) and d is the annual discount rate. Using discount factors Equation 4.19 can be written as

$$\text{NPV} = \sum_{n=0}^{N} F_n \text{DF}_n = F_0 + F_1 \text{DF}_1 + F_2 \text{DF}_2 + \cdots + F_N \text{DF}_N \quad (4.20)$$

where DF_n is the discount factor for year n. In the simple case where projects have an initial capital investment in year 0, subsequent equal annual undiscounted net operating cash flows over the investment lifespan and no residual value, Equation 4.20 can be simplified to

$$\text{NPV} = F_{c,0} + F_{no} \sum_{n=0}^{N} DF_n = F_{c,0} + F_{no}(\text{DF}_1 + \text{DF}_2 + \cdots + \text{DF}_N) \quad (4.21)$$

where $F_{c,0}$ is the capital investment cash flow which occurs in year 0 and F_{no} is the net annual operating cash flow which is the same for all years.

Instead of summing all discount factors over the life of a project, a CDF can be used. CDFs for different time periods and discount rates are provided in Appendix A. Equation 4.21 can then be written as

$$\text{NPV} = F_{c,0} + F_{no} \text{CDF}_{N,d} \quad (4.22)$$

where $\text{CDF}_{N,d}$ is the CDF for a lifespan of N years at a discount rate d.

Example 4.17   Net Present Value

A utility is investigating the economic feasibility of investing in a 140 MW compressed air energy storage (CAES) plant with an estimated capital cost of €155m. Fuel, compression and O&M costs are projected to be €18m/annum, while revenues from energy sales and system services are expected to be €38m/annum. The project has a lifespan of 30 years at the end of which the project has an estimated salvage value of €22m. The company has a

WACC of 6% (real). Calculate the NPV for the project using the following approaches:

1. using the discount rate approach assuming an overnight build and decommissioning;
2. using a CDF assuming an overnight build but no decommissioning cost; and
3. using the discount rate approach assuming that the project will take three years to build with capital expenditures of €32m, €32m and €91m in years 1, 2 and 3 respectively, and where decommissioning will take 1 year to complete.

The solutions to these problems are given as follows.

1. *Discount rate Approach with Overnight Build and Decommissioning.* Table 4.12 shows the annual cash flows for the project which are discounted using the 6% WACC discount rate. The discounted cash flows are summed over the project lifespan, as outlined in Equation 4.19, to give a net present value of €124m.

2. *Cumulative Discount Rate Approach Ignoring Decommissioning Costs.* For a discount rate of 6% and an investment period of 30 years, the discount factor table in Appendix A gives a CDF of 13.765. Substituting an annual net operating cash flow of €20m and an initial investment cash flow of €155 into Equation 4.22, we get

$$NPV = -155 + 13.675 \times 20 = €120m$$

3. *Discount Rate Approach with 3-Year Build and 1-Year Decommissioning.* Table 4.13 shows the annual cash flows for the project showing the extended construction and decommissioning periods. The 30-year operational period occurs in project years 4–33. An NPV of €90m is obtained, lower than the €124m for the overnight build because positive cash flows have been pushed further into the future and therefore have a lower present value.

Table 4.12   NPV calculations for the CAES project using a discount rate of 6% assuming an overnight build (all figures in €m).

					Year						
	0	1	2	3	4	5	...	27	28	29	30
Capital investment/residual value	(155)										22
Operational costs		(18)	(18)	(18)	(18)	(18)	...	(18)	(18)	(18)	(18)
Revenues		38	38	38	38	38	...	38	38	38	38
Net cash flow	(155)	20	20	20	20	20	...	20	20	20	42
Discounted cash flow (6%)	(155)	19	18	17	16	15	...	4	4	4	7
Net present value	124										

Selected years only are shown.

Table 4.13 NPV calculations for the CAES project using a discount rate of 6% assuming a 3-year build and 1-year decommissioning (all figures in €m).

	Year											
	1	2	3	4	5	6	...	28	29	30	31	32
Capital investment/residual value	(32)	(32)	(91)									
Operational costs				(18)	(18)	(18)	...	(18)	(18)	(18)	(18)	(18)
Revenues				38	38	38	...	38	38	38	38	38
Net cash flow	(32)	(32)	(91)	20	20	20	...	20	20	20	20	20
Discounted cash flow (6%)	(30)	(28)	(76)	16	15	14	...	4	4	3	3	3
Net present value	90											

Selected years only are shown.

NPV is widely regarded as the most representative measure of the financial performance of an investment and is used in all industry sectors on all types of projects as it considers both the time value of money and investment size. It can be used in a range of different decision types including mutually exclusive projects, projects of different sizes, since it differentiates between investment sizes, and projects involving social costs (dealt with in Chapter 5). It is should be used wherever possible as a primary measure of financial performance.

NPV does have weaknesses. It assumes that additional cash flows generated by the project are reinvested at the discount rate over its lifespan, which is unlikely to be the case. The calculation of NPV is similar to compound earnings in a savings bank account, where real interest earnings are retained and reinvested with the original sum to earn interest. However, the additional cash earned for an energy project cannot normally be reinvested in that project because it has already been built; therefore, in reality, it has to be reinvested elsewhere and often at a different discount rate. This weakness of NPV can be overcome using a modified NPV, which allows for different discount and reinvestment rates (for further information see Clarke et al. (1989)).

A second weakness is that NPV assumes an unvarying discount rate over the lifespan of the project. For example, a company's funding structure may vary over time and it may, therefore, need to assess projects using varying discount rates. Where information on different required discount rates is available, the NPV can be calculated using

$$NPV = \sum_{n=0}^{N} \frac{F_n}{\prod_{i=1}^{N}(1+d_i)^n} \tag{4.23}$$

where d_i is the discount rate for period i and Π is the product operator.

We have discussed that it can be difficult to choose an appropriate discount rate which represents a company's or person's required rate of return. One option to address this difficulty is the generation of a project NPV profile for different discount rates which graphically represents the NPV of the project for different discount rates.

Example 4.18   NPV Profile

A renewable energy project which requires an investment of €8m will yield annual net operating cash flows of €2m. The investor has no history of investing in this type of project and does not know what discount rate to use in order to assess its attractiveness. Therefore, an NPV profile is created for the project using Equation 4.22 by calculating the project NPV for different discount rates varying from 0% to 50%. Results are show in Figure 4.2 and Table 4.14. The investor can proceed knowing that the project must achieve a real WACC of less than 24.7% in order for it to be viable.

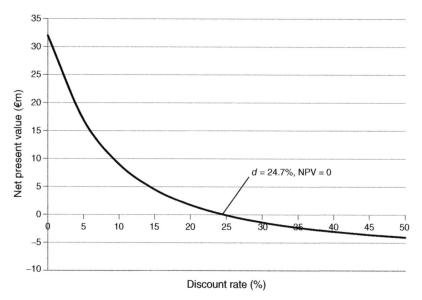

Figure 4.2   A plot of the NPV results from Table 4.3 showing a discount rate intercept value (an NPV of zero) of 24.7%.

4.3.5   Internal Rate of Return

The IRR is the discount rate at which a project NPV is equal to zero. Under this condition, the sum of discounted net cash inflows equals the sum of the net cash outflows (including investment cash flows) over the lifespan of the project. This can be expressed as

$$\sum_{n=0}^{N} \frac{F_{i,n}}{(1+\mathrm{irr})^n} = \sum_{n-0}^{N} \frac{F_{o,n} + F_{c,n}}{(1+\mathrm{irr})^n}$$

(4.24)

Table 4.14 Investment and net cash flows are shown for the energy project together with the NPV for a range of different discount rates (real cash flows and discount rates are used).

Discount rate (%)	Investment cost (€m)	Net annual cash flow (€m)	Cumulative discount factor	Cumulative discounted net cash flow (€m)	Net present value (€m)
0	8	2	20.00	40.00	32.00
5	8	2	12.46	24.92	16.92
10	8	2	8.51	17.03	9.03
15	8	2	6.26	12.52	4.52
20	8	2	4.87	9.74	1.74
25	8	2	3.95	7.91	−0.09
30	8	2	3.32	6.63	−1.37
35	8	2	2.85	5.70	−2.30
40	8	2	2.50	4.99	−3.01
45	8	2	2.22	4.44	−3.56
50	8	2	2.00	4.00	−4.00

where $F_{i,n}$ and $F_{o,n}$ are the sums of cash inflows and outflows, respectively, in year n, $F_{c,n}$ is the investment cash flow in year n, N is the investment lifespan and irr is the discount rate which satisfies equality. The equation is normally solved iteratively using a computer.

Example 4.19   Internal Rate of Return

A company is considering placing 300 kWp of PV panels on the roof of its office building at an estimated cost of €630,000. The system is projected to produce 268.8 MWh of electricity annually which would benefit from a long-term (25 years) power purchase agreement (PPA) at a rate of 28 c/kWh giving estimated annual revenues of €75,264. Incremental operating costs are estimated to be negligible and are ignored. Does the investment cover the company's cost of capital of 15%?

Substituting an investment cash flow of €630,000 and operating cash inflows of €75,264 into Equation 4.24 we get

$$\sum_{n=1}^{25} \frac{75{,}264}{(1+\mathrm{irr})^n} = \frac{630{,}000}{(1+\mathrm{irr})^0} \text{ so that irr} = 11.1\%$$

Solving this iteratively by hand or, more commonly by calculator or spread sheet function, the IRR is found to be 11.1%, which does not meet the company's MARR of 15%.

IRR is a convenient way of comparing project returns to other investment types such as bonds and other financial investments and, for this reason, it is a popular measure with managers and decision makers. However, it suffers from a number of significant drawbacks. It does not differentiate between investment sizes and is therefore not appropriate for choosing between mutually exclusive projects of different investment scales. For example, a very small project with a high IRR may be favoured over a larger one with a lower IRR even though the latter may be the most valuable way to invest the sum available to the investors. A second problem with IRR is that a project with the highest IRR may not necessarily provide the highest NPV at the investor's MARR. This is illustrated in Figure 4.3, which shows NPV profiles for two different energy projects. For an investor with a MARR of 9.7% or less, the NPV of project B is greater although its IRR is lower. Above this threshold, the IRR and NPV give the same ranking.

IRR also suffers from the same problem as NPV where it assumes that additional cash flows generated by the project are reinvested at the IRR over its lifespan. This problem is, however, normally more exaggerated in the case of IRR, because WACC is normally used for NPV calculations and there are normally opportunities in the company to reinvest surplus project cash flows close to this rate. However, because IRR results in an NPV of 0, it is often higher than the WACC, and therefore, earnings on surplus cash flows are exaggerated. The problem is overcome using the modified internal rate of return (MIRR). The true rate of return is a combination of the IRR and reinvestment rate earned on surplus cash flows. MIRR assumes that positive cash flows are reinvested

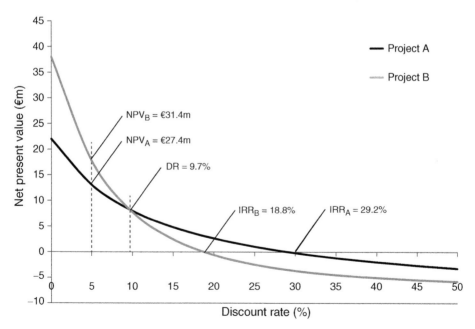

Figure 4.3   NPV profiles for two projects with different cash flow profiles showing how a project with the highest IRR may not have the highest NPV. Below a minimum acceptable rate of return of 9.7% IRR selects the less valuable project.

at a rate which is achievable by the company (normally the WACC) for the duration of the project; the sum of these cash flows is then discounted to its present value. The MIRR is the discount rate that equates this value with the present value of all negative cash flows discounted to today's value using the project finance rate. MIRR is given by

$$\sum_{n=0}^{N} \frac{Fn_n}{(1+d_f)^n} = \frac{\sum_{n=0}^{N} Fp_n(1+d_r)^{N-n}}{(1+\text{mirr})^N} \tag{4.25}$$

$$\text{mirr} = \sqrt[N]{\frac{\sum_{n=0}^{N} Fp_n(1+d_r)^{N-n}}{\sum_{n=0}^{N} \frac{Fn_n}{(1+d_f)^n}}} - 1 \tag{4.26}$$

$$\text{mirr} = \sqrt[N]{\frac{\text{Future value of positive cash flows at reinvestment rate}}{\text{Present value of negative cash flows at finance rate}}} - 1 \tag{4.27}$$

d_r is the reinvestment rate, d_f is the finance rate and mirr is the modified internal rate of return.

Another failing of IRR is that projects with positive net cash flows followed by negative ones (as opposed to the typical situation where negative upfront investment costs are followed by positive net cash flows) may have multiple IRRs. This may occur in phased energy projects where there are significant additional investment costs some years after project start-up. MIRR overcomes this problem.

Example 4.20   Modified Internal Rate of Return

The company referred to in Example 4.19 wishes to estimate the MIRR of the PV project. The project can be financed at a rate of 10%, and it is assumed that any cash generated can be reinvested at the company's WACC of 15%. Using these finance and reinvestment rates, together with the investment cash flow of €630,000 and annual operating cash inflows of €75,264 over 25 years in Equation 4.26 we get

$$\text{MIRR} = \sqrt[25]{\frac{\sum_{n=0}^{25} Fp_n(1+0.15)^{25-0}}{630,000/(1+0.1)^0}} - 1 = \sqrt[25]{\frac{16,015,654}{630,000}} - 1 = 13.8\%$$

The MIRR (13.8%) is higher than the IRR (11.1%) because the positive cash flows from the project are invested at the company's cost of capital (15%) rather than at the IRR (11.1%) and because the project is financed at a rate (10%) which is lower than the IRR. However, it still does not meet the company's 15% threshold.

4.3.6   Life cycle cost

The LCC of a project is the sum of all discounted costs over its lifespan. Costs include cash outflows relating to investment, fuel, operation, maintenance, overheads and other costs of doing business. Capital subsidies and residual values are included, but LCC ignores all recurrent cash inflows such as energy sales and avoided energy costs. It is given by

$$\text{LCC} = \frac{\sum_{n=0}^{N}(F_{c,n} + F_{o,n})}{(1 + d)^n} \tag{4.28}$$

where LCC is the present value of the life cycle cost, $F_{c,n}$ is the investment cost in period n, $F_{o,n}$ is the cash outflows in period n, N is the total number of periods and d is the discount rate.

Example 4.21   Life Cycle Cost

A 600 MW combined cycle gas turbine (CCGT) power station is being built at a cost of €0.5m/MW. It is projected that the plant will have a capacity factor of 0.6 operating at an average efficiency of 57% over 40 years. The main costs include operation and maintenance costs at 2.80 €/MWh and gas costs at 12.6 €/MWh. If inflation is 2% and the nominal discount rate is 10%, calculate the project LCC ignoring taxes and project residual value.

It is estimated that annually the plant produces 3154 GWh(e) (24 h × 365 days × 0.6 × 600 MW) and consumes 5533 GWh of gas (3154 GWh(e)/0.57 efficiency). Cash flows are calculated as follows:

Investment costs = Installed capacity × Unit cost = 600 × 0.5 = €300m

Operating cash outflows

   = Electricity ouput × O&M tariff + Gas consumed × Gas tariff

   = 3154 × 2.8 + 5533 × 12.6

   = €78.5 m/annum

These cash outflows are first inflated at 2% per annum to give nominal prices for each year of operation and then discounted at the 10% nominal discount rate to give the present value of total costs for each period. Using

Equation 4.28, the sum of these discounted annual costs is €1252.5, which is the LCC of the project. Results are presented in Table 4.15.

Table 4.15 Life cycle costs (in €m) for a 600 MW CCGT plant.

							Year						
	0	1	2	3	4	5	...	35	36	37	38	39	40	LCC
Investments costs	300.0												
Real operating costs		78.5	78.5	78.5	78.5	78.5	...	78.5	78.5	78.5	78.5	78.5	78.5
Total real costs	300.0	78.5	78.5	78.5	78.5	78.5	...	78.5	78.5	78.5	78.5	78.5	78.5
Total nominal costs	300.0	80.1	81.7	83.3	85.0	86.7	...	157.1	160.2	163.4	166.7	170.0	173.4
Total discounted costs	300.0	72.8	67.5	62.6	58.1	53.8	...	5.6	5.2	4.8	4.5	4.1	3.8	1252.5

Selected years only are shown.

LCC is used to rank projects where the benefits are identical. Take, for example, the case where an off grid battery needs continuous charging (such as a motorway 'SOS' phone) using either micro-wind or PV technologies. Here, the LCCs of the PV and micro-wind turbines can be directly compared as the benefits (electrical energy supplied) are identical in each case. Similarly, LCC can be used to evaluate the effects of different timings of costs between competing alternatives. For example, where there is an option to procure the same wind farm in two different ways, each with different phased payments, LCC could be used to compare alternatives, assuming that cash inflows (benefits) are identical in magnitude and timing.

LCC is useful for comparing projects where a certain level of service must be provided and the value of the benefits are difficult to determine. For example, it could be used to assess the relative costs of different systems for providing cooling to an office building.

4.3.7 Levelised Cost of Energy

The LCOE is the cost that, if assigned equally to every unit of energy produced (or saved) by the system over the project lifespan, will equal the LCC when discounted back to the base year. It represents the present value of the energy tariff that must be charged over the project lifespan for the investment to have an NPV of 0. In this breakeven situation, the total discounted cash outflows and inflows are equal. Therefore, the sum of discounted cash outflows must equal the sum of discounted cash inflows:

$$\frac{\sum_{n=0}^{N}(F_{c,n} + F_{o,n})}{(1+d)^n} = \frac{\sum_{n=0}^{N}(F_{i,n})}{(1+d)^n} \tag{4.29}$$

where $F_{c,n}$ is the investment cost in period n, $F_{o,n}$ is the cash outflows in period n, $F_{i,n}$ are the cash inflows in period n, N is the total number of periods and d is the discount rate.

As the term on the left of the equation is LCC (see Equation 4.28), and that on the right is the product sum of energy output and the energy sales tariff, we can write

$$LCC = \frac{\sum_{n=0}^{N}(E_n \times t)}{(1+d)^n} \tag{4.30}$$

where E_n is the energy sold by the project in year n and t is the energy sales tariff for each unit of energy. However, in the special breakeven case described, the energy sales tariff is the LCOE. Rearranging we find that the LCOE is the LCC divided by the energy output discounted to the base year:

$$LCOE = \frac{LCC}{\sum_{n=1}^{N}[E_n/(1+d)^n]} \tag{4.31}$$

LCOE is useful for comparing the unit energy costs of diverse energy technologies operating at different scales and over different time periods. In this way, it can be used to rank the attractiveness of alternative energy investments assuming that the cash inflows from the energy produced by each are identical. However, a technology which produces more electricity during cheaper, off-peak periods cannot be compared using LCOE to one which produces the same amount of electricity at more valuable peak periods, as revenues will be different in both cases. LCOE cannot be used to compare different forms (e.g. heat and electricity) of energy as these provide different benefits. LCOE cannot be used to select among mutually exclusive alternatives because different investment sizes are not considered.

Example 4.22   Levelised Cost of Energy

Calculate the LCOE for Example 4.21.

Equation 4.31 requires the LCC and the sum of electricity output to be discounted to today's worth. We know that the LCC for Example 4.21 is €1253.2m. In order to calculate the discounted energy outputs, we first calculate the real discount rate using Equation 4.8 with an inflation rate of 2% and a nominal discount rate of 10%:

$$d_r = \left[\frac{(1+0.10)}{(1+0.02)}\right] - 1 = 7.8\%$$

The LCOE is then calculated using Equation 4.31, taking an annual electricity output from the plant of 3156 MWh over its 40-year lifespan to give an

LCOE of 3.27 c/kWh:

$$\text{LCOE} = \frac{1253.2 \times 10^6}{\displaystyle\sum_{n=1}^{40}[3156/(1+0.078)^n]} = \text{€}32.70/\text{MWh}$$

4.3.8 Uncertainty and risk

Uncertainty about inputs to energy project cash flow and financial models results in the risk of the actual financial performance deviating from that which is projected. This risk can be positive or negative resulting in better-than- or worse-than-expected financial performances, respectively. For example, uncertainty about future biomass boiler fuel prices results in the risk of lower profits if prices are high or greater profits if they are low. Uncertainty can be classified in a wide variety of different ways; however, two broad types are well recognised: *aleatory*, which is due to inherent randomness, and *epistemic*, which is due to incomplete measurement or inaccurate modelling of a phenomenon. An example of the former might be annual variations in average wind speeds used in wind farm investment modelling. An example of the latter could be lack of accurate information on the company's WACC. Epistemic uncertainty can be reduced by further measurements or better modelling (subject to the benefits of the accurate data outweighing collection costs) whereas aleatory uncertainty can only be quantified. For energy investment projects, the greatest uncertainties typically relate to the future values of key inputs over the investment lifespan. For example, when assessing an investment in a grid-connected PV system there will be a very large uncertainty about the price of grid electricity in 25 years' time.

There is no energy investment project which is without uncertainty and risk, no matter how much effort is put into risk management and mitigation. However, it is important to fully consider risk in any investment decision because a project with large downside risks may not fit the risk appetite of the investing organisation or individuals. Furthermore, if the project proceeds, then a good understanding of the sources of risk facilitates better project risk management. There are a number of approaches which can be used to quantify investment risk:

1. *Sensitivity analysis* which quantifies the changes in the chosen financial parameters for changes in inputs. For example, an increase of 10% in the capital cost of a wind turbine might result in a 3% reduction in its NPV or a 10% increase in average wind speed may result in a 5% increase in NPV.
2. *Scenario analysis* involves identifying likely future 'worlds' and choosing consistent associated technical and financial inputs, which result in a different financial assessment for each scenario. It differs to sensitivity analysis in that multiple input parameters are changed simultaneously rather

than individually. For example, a wave energy project to be implemented in 5 years may look at scenarios of 'high energy prices and rapid technology development' and 'low energy prices and slow technology development' because wave energy technology development is likely to be spurred by high energy prices. Input values for energy sales and capital costs which are consistent with these future worlds are then chosen and the financial performances for both scenarios are established.

3. *Monte Carlo simulation* involves estimating input probability distributions, repeatedly randomly sampling each of these distributions to give a distribution of outputs. For example, a 25-year model of the PP for a domestic PV system is a function of system cost, annual solar irradiance, system efficiency and the price of displaced electricity. Input distributions are estimated for capital cost, irradiation and electricity price to give the PP probability distribution for the system. System efficiency may not be subject to sufficient uncertainty to be included as an input probability distribution.

Example 4.23   Project Risk – Sensitivity Analysis

A company is considering investing in an energy project which will produce 1.5 GWh of electricity per annum which will be sold at 4 c/kWh over 15 years. Estimates of capital and O&M costs are €250,000 and €20,000 per annum, respectively. The company's WACC is 6% (real). An overnight build is assumed and depreciation and taxation are ignored. Undertake a sensitivity analysis for the project using NPV as the financial measure preferred by management.

The NPV is calculated for the 'base case' using the estimated values above, as shown in Table 4.16. Uncertainty analysis of input values suggests that the electricity tariff, O&M costs and capital costs are uncertain but that electricity production and WACC are sufficiently certain not to be considered further as probabilistic inputs. A view is taken that the electricity tariff could vary by ±30%, O&M cost by −10% to +30% and capital cost from 0% to +30%.

Table 4.16   NPV for the energy investment showing cash inflows and outflows (monetary values in €000s).

								Year								
	0	1	2	3	4	5	6	7	8	9	10	11	12	13	14	15
Cash out	(250)	(20)	(20)	(20)	(20)	(20)	(20)	(20)	(20)	(20)	(20)	(20)	(20)	(20)	(20)	(20)
Cash in		60	60	60	60	60	60	60	60	60	60	60	60	60	60	60
Net cash	(250)	40	40	40	40	40	40	40	40	40	40	40	40	40	40	40
DF	1.00	0.94	0.89	0.84	0.79	0.75	0.70	0.67	0.63	0.59	0.56	0.53	0.50	0.47	0.44	0.42
PV	(250)	38	36	34	32	30	28	27	25	24	22	21	20	19	18	17
NPV	138															

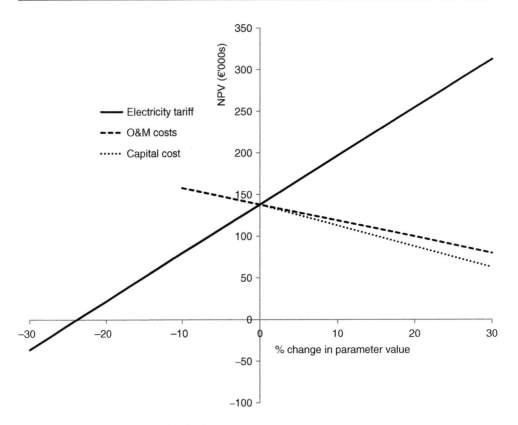

Figure 4.4   Sensitivity results for the investment.

The value of each input parameter is then varied in increments of ±10% up to the relevant threshold values and the resultant NPV recorded. Figure 4.4 shows a plot of NPV against percentage changes in the three uncertain parameter values. It can be seen that electricity tariff results in the greatest variation in NPV (it has the largest slope) and that a decrease of more than 25% results in a negative NPV. Management may, therefore, opt to mitigate or eliminate this risk, perhaps by entering into a long-term PPA at a suitable tariff. Capital and O&M costs are not expected to result in negative NPVs so are not managed as closely. A more complete analysis would consider combinations of variations in input parameters (scenario analysis) and/or Monte Carlo analysis.

4.3.9   Financial measures compared

The strengths and weaknesses of the various financial parameters described earlier along with the suitability of different decision types are summarised in Table 4.17.

Table 4.17 A summary of the strengths and weaknesses of the different financial parameters.

Financial parameter	Strengths	Weaknesses	Accept/reject	Mutually exclusive projects	Ranking
Payback period	Simple to calculate. Measures the liquidity of the project. Measures how long investment is at risk	Does not quantify project value. Cash flows after payback are ignored. Time value of money not considered	Can be used	May be wrong	May be wrong
Discounted payback periods	Simple to calculate. Measures the liquidity of the project. Measures how long investment is at risk. Considers time value of money	Does not measure project value. Cash flows after payback are ignored	Can be used	May be wrong	May be wrong
Return on investment	Simple to calculate. Widely used so can be easily compared to other investments	Does not quantify the value of an investment, only gives a % return. Time value of money not considered	Can be used	May be wrong	May be wrong
Profitability index	Measures whether benefits outweighs costs. Considers the time value of money. Includes all cash flows	Does not quantify the value of an investment, only gives a ratio	Can be used	May be wrong	Recommended
Savings-to-investment ratio	Measures whether benefits outweighs costs. Considers the time value of money. Includes all cash flows	Does not quantify the value of an investment, only gives a ratio	Can be used	May be wrong	Recommended
Net present value	Quantifies the value of an investment. Considers the time value of money. Includes all cash flows	Assumes that surplus cash flows reinvested at the discount rate	Recommended	Recommended	Recommended

(continued overleaf)

Table 4.17 (*continued*)

Financial parameter	Strengths	Weaknesses	Accept/reject	Mutually exclusive projects	Ranking
Internal rate of return	Indicates whether investment increases value. Considers the time value of money. Includes all cash flows. Easy to compare to other investment options	Does not quantify the value of an investment, only gives a % return. Assumes positive cash flows reinvested at IRR. Cannot be used when sign of net cash flows changes more than once	Can be used	May be wrong	May be wrong
Modified internal rate of return	Indicates whether investment increases value. Considers the time value of money. Includes all cash flows. Easy to compare to other investment options	Does not quantify the value of an investment, only gives a % return	Can be used	May be wrong	May be wrong
Life cycle cost	Considers the time value of money	Can be used to compare projects with equal benefits only. Includes all investment cash flows only – ignores cash inflows. Does not quantify the value of an investment	No	Can be used	Can be used
Levelised cost of energy	Considers time value of money. Useful for comparing unit costs of different investments	Includes all investment cash flows only – ignores cash inflows. Does not quantify the value of an investment	No	May be wrong	Can be used

NPV is typically recommended as a primary measure supported by other appropriate measures.

4.4 Case studies

4.4.1 Municipal bus fleet conversion to compressed natural gas

A small regional city has a publically owned municipal bus company serving urban commuters. The company must replace its fleet over the next 5 years and is considering two options:

1. replacement with diesel buses similar to those which it currently operates; or
2. replacement with compressed natural gas (CNG) buses which are cleaner and quieter but will require substantial up-front fuelling infrastructural investment at the bus depot.

CNG bus technology is almost identical to the more common diesel technology. Diesel engines require conversion which has an additional cost. The CNG fuel is predominantly methane (CH_4) which when burned with oxygen forms carbon dioxide and water. Because it has a high energy-to-carbon ratio, it produces less CO_2 compared to other fuels, such as diesel. It also produces less harmful pollutants such as PM2.5s and PM10s (particles smaller than 2.5 and 10 μm, respectively) which are particularly damaging to human health.

Bus depots must be modified to cater for CNG vehicles. The refuelling technology requires a gas grid connection, compressors, storage and dispensers. The maintenance costs of operating CNG buses are similar, while savings can be realised due to the lower cost of natural gas over diesel.

The bus company therefore wants to determine whether the additional depot upgrade and vehicle costs are offset by the financial benefits of lower fuel costs.

The first step in the financial study is to identify the business-as-usual (BAU) case so that the incremental costs of the CNG project can be clearly identified. The current plans of the company are to replace its fleet of 45 buses over the next 5 years as shown in the schedule in Table 4.18. This involves the purchase of new diesel buses, each costing €225,000 with a fuel efficiency of 45 km/100 l and maintenance costs of €8530 per annum. All costs are given in 2014 prices, which are chosen as the base year for the study. The company knows that each bus travels on average 85,000 km each year and it is

Table 4.18 Vehicle replacement schedule for the bus company.

	Year				
	1	2	3	4	5
Number of buses purchased	7	10	14	8	6
Cumulative number of buses	7	17	31	39	45

anticipated that this distance will not change over the duration of the project. Buses last for approximately 10 years at which stage they are scrapped with no net value. Currently, the company pays €1.02/l of diesel excluding value-added taxes and other reclaimable fuel taxes. Diesel refuelling facilities are currently available at its depot.

Each of the alternative CNG vehicles costs €275,000 and achieves a fuel efficiency of 42 km/100 l of diesel equivalent (i.e. the amount of gas with the same energy content as 1 l of diesel). Their maintenance cost is projected at €8530 per vehicle per annum over a 10-year service life (again, scrappage costs are negligible). Enquiries with natural gas suppliers indicate that gas will cost 3.5 ¢/kWh. Preliminary designs and costs have been obtained for the required CNG refuelling station and associated depot modifications; estimated costs are €1.8m and €750,000, respectively.

Given that the energy density of the diesel used in the buses is 9.96 kWh/l and a 3.5 ¢ unit cost of gas, the CNG cost per litre of diesel equivalent is €0.35 (9.96 × 3.5 ÷ 100), approximately one-third of the cost of a litre of diesel (Table 4.19).

Because the project is publically funded, the discount rate recommended by the government's Department of Finance is used. This stipulates a 5.5% nominal discount rate for projects lasting between 10 and 20 years; this project will be assessed over 15 years (the last bus will be purchased in year 5, and it will last for a further 10 years). A medium to long term HICP rate of 2% is recommended for general costs, but where this is not likely to be reflective of certain cost components other indices may be used. In the case of the CNG project, gas and diesel price trends were further investigated given their central importance and the globalised nature of these commodities. The US Department of Energy 'Annual Energy Outlook 2014' (AEO2014) was used to estimate future fuel price inflation using their central reference case scenarios. Gas price nominal inflation is anticipated to be 3.7% over the medium term (up to 2040). Crude oil price inflation was used as a proxy for diesel price inflation since it is the greatest cost input; this is anticipated to be 2.9% over the medium term. Although gas price inflation is higher than that of oil, gas prices are starting at a historically low level. Inflation indices are summarised in Table 4.20.

Two financial parameters are used in this assessment: NPV, which measures the total value of the investment to the company, and SIR, which indicates the relative ratio of discounted savings to the required capital investment. A

Table 4.19 Fuel unit costs used in the CNG project assessment.

Fuel	Price	Units
Diesel price (2014)	1.02	€/l
Gas price (2014)	3.50	cent/kWh
Gas price (2014)	0.35	€/equivalent litre of diesel

Table 4.20   Inflation indices used in the
financial appraisal.

Inflation index	Rate per annum (%)
HICP	2.0
Diesel price inflation	2.9
Gas price inflation	3.7

sensitivity analysis is also undertaken to assess the impacts of uncertain inputs on these financial parameters. The impacts of 10% and 20% increases in capital costs, oil price inflation and gas price inflation are considered in this regard.

The first step in the assessment process is to determine the incremental costs and benefits of the CNG project over the BAU diesel option. Table 4.21 shows the costs of the BAU and CNG projects as well as the incremental cost of the CNG project over the BAU case. Items with the same costs for both projects such as insurances and site rental are ignored since these do not result in incremental costs or benefits. It can be seen that the incremental capital costs of building the refuelling station and modifying the depot are €1.80m and €0.75m, respectively. The additional capital and maintenance costs for each bus are €50,000 (once off) and €2320 per annum, respectively. Fuel savings per bus are €26,570 per annum.

Once these incremental costs and benefits are identified, nominal cash flows are calculated as shown in Table 4.22. Bus purchases are calculated for each year based on the schedule in Table 4.18, incremental bus purchase costs (€50,000) and the HICP of 2% per annum. Incremental bus maintenance cash flows are similarly calculated. Incremental bus fuel costs are calculated as the difference between fleet diesel and CNG costs for that year, each adjusted for diesel and natural gas inflation rates, respectively (as listed in Table 4.20). All of these costs and benefits are summed to give the total incremental cash

Table 4.21   Project option costs and incremental costs and benefits (negative costs in parenthesis).

Item	BAU diesel option	CNG option	Incremental benefit (cost)	Units
Fuelling station cost	0	1.8	(1.80)	€m
Depot modification	0	1	(0.75)	€m
Bus purchase cost	2,25,000	2,75,000	(50,000)	€/bus
Maintenance	8,530	10,850	(2,320)	€/bus/annum
Fuel efficiency	45	42		litres/100 km diesel or equivalent
Annual mileage	85,000	85,000		km/annum
Annual fuel consumption	38,250	35,700		litres diesel or equivalent/annum
Annual fuel cost	39,015	12,445	26,570	€/bus/annum

Table 4.22 Incremental nominal cash flows for the compressed natural gas (CNG) bus project compared to the business as usual (BAU) diesel bus alterative.

	Year															
	0	1	2	3	4	5	6	7	8	9	10	11	12	13	14	15
Fuelling station installation	(1800)															
Depot modifications	(750)															
Bus purchases		(357)	(520)	(743)	(433)	(331)										
Bus maintenance		(17)	(41)	(76)	(98)	(115)	(118)	(120)	(122)	(125)	(127)	(130)	(132)	(135)	(138)	(141)
Bus fuel costs		191	475	888	1145	1354	1388	1422	1458	1494	1531	1569	1608	1648	1688	1730
Total incremental cash flows	(2550)	(183)	(86)	68	614	907	1,270	1,303	1,336	1,369	1,404	1,439	1,476	1,513	1,551	1,589
Total discounted incremental cash flows	(2550)	(173)	(78)	58	495	694	921	895	870	846	822	799	776	754	733	712
Discount Factor		*1.000*	*0.948*	*0.898*	*0.852*	*0.807*	*0.765*	*0.725*	*0.687*	*0.652*	*0.618*	*0.585*	*0.555*	*0.526*	*0.499*	*0.473*	*0.448*

Table 4.23 Sensitivity of project NPV (€m) to variations in key parameter input values.

Increase in original value	Capital cost	Oil price inflation	Gas price inflation
None	6575	6575	6575
10%	6395	7025	6378
20%	6215	7487	6173

flows, which are then converted to total discounted incremental cash flows using the relevant nominal discount factor (based on the nominal discount rate of 5.5%).

Summing the discounted cash flows gives a positive NPV of €6.575m and an SIR of 3.6, both indicating that the CNG project is worth investing in. Sensitivity analysis results (see Table 4.23) show that the NPV of the project is not significantly exposed to the modelled changes in input parameter values.

4.4.2   New wind farm development

A wind farm developer is considering investing in a new wind farm and wishes to assess its financial performance. The project is located on a windy site on the Atlantic coast, which has planning permission for two turbines with hub heights of 85 m and a connection agreement to the electricity grid. The company has an option of a 20-year land lease, extendable to 25 years at a rate of 3% of all revenues (cash inflows). The wind resource at the site has been assessed over 1 year by a reputable wind energy consultancy firm, and a PPA is available from an energy supply company to purchase all electrical output from the site at a fixed rate for a 15-year period. The company wants to establish if the project is worth investing in.

A wind resource site survey was undertaken for 1 year using a monitoring mast and results were statistically adjusted to reflect the expected long-term wind resource on the site. The study indicates the following wind resource on the site

- a P50 net energy yield of 14,500 MWh/annum and
- a P90 net energy yield of 12,250 MWh/annum.

A P50 or P90 yield is a measure of the probability that the site will meet or exceed this wind yield. For this site, therefore, there is a 90% probability that the energy yield will be at least 12,250 MWh/annum and a 50% chance that it will produce more (or less) than 14,500 MWh/annum. For the investors, who are exposed equally to upside and downside risks, it is most appropriate to use the P50 metric. Banks, however, usually take a more conservative view as they are more exposed to the downside risks. For example, if the project goes extremely well they will only get repaid the loan and interest, whereas if it goes badly they are left with a bad debt; they tend to use the P90 figure in their assessments. A P50 figure will be used here since the financial appraisal is being undertaken on behalf of the investors. Sensitivity analysis will be used to assess risk.

An engineering consultancy is employed to develop a design and cost plan for the proposed development. The resulting capital and O&M cost estimates are presented in Tables 4.24 and 4.25. The biggest cost component is the wind turbine equipment costs at €4.2m. A lease premium is paid to the property owner upon signing of the lease for the wind farm site; wayleaves must be obtained from landowners for power lines and other infrastructure. Together, the lease and wayleaves cost €1.225m. The finance cost is the cost of funding the construction period of the investment before operation commences and the 15-year loan agreed with the bank comes into operation; this costs €280,000. Other costs bring the total capital cost of the wind farm to €6.665m. The biggest O&M cost is for the full maintenance of the wind turbines, equipment and site. It is a requirement of the bank that a 15-year maintenance

Table 4.24   Capital costs for the wind farm development.

Item	Cost (€000s)
Equipment costs	4200
Lease premium and wayleaves	1225
Civil engineering and structures	105
Grid connection and TSO payments	545
Professional fees, insurances, other costs	200
Development finance costs	280
Contingency	110
Total	*6665*

Table 4.25   Operating and maintenance cost estimates (in €'000s) for the project.

Rent	3% of cash inflows
Rates	€22 per annum
Directors fees	€10 per annum
Management charges	€130 per annum
Insurance	€27 per annum
Accounting fees	€2 per annum
Audit	€3 per annum
General expenses	€3 per annum

contract is signed with a reputable wind farm maintenance contractor to protect the investment over the loan period.

The wind farm developer has the option of becoming a registered energy supplier and selling all electricity output onto the 'spot' wholesale market or by entering into a PPA with a registered energy supplier who will pay a fixed price for the electricity obtained. Given that the project is highly exposed to fluctuations in the price of electricity (see sensitivity analysis below), the wind farm development company opts to enter into a PPA with a large energy supplier. A price of 85 €/MWh is agreed for a contract period of 15 years; again this period is needed if the bank is to provide a 15-year loan to the project. The revenues from the project are therefore the product of the P50 energy yield and the PPA price:

$$\text{Revenues (Cash inflows)} = 14{,}500 \times 85 = 1.220\text{m}€/\text{annum}$$

Depreciation is calculated on a straight-line basis, with the book value decreasing by 5% of the original capital cost per annum over 20 years. Corporation tax is 20% of earnings after accounting for depreciation. There are no capital subsidies or other tax incentives available for the project.

The wind farm development company has presented the project to the banks who have offered 80% funding at a loan rate of 6% over 15 years. Returns on equity in the company are historically 14%, thus giving a WACC of 7.6% (real).

Three financial parameters are chosen for assessing the project:

1. NPV because it provides the best measure of how much value the project will add to the wind development company;
2. IRR because the investors can easily compare this to returns on other investments; and
3. The LCOE because this can be easily compared to PPA unit revenues.

Sensitivity analysis is used to assess the effects of input parameter uncertainties on the project's financial viability using NPV as the most important financial parameter. Two approaches are taken: the first is to vary all key input parameters to identify their effects on NPV so that the development company can better understand and manage project risk. The second is to investigate scenarios of particular interest to the company which include

1. 'Low Wind', where wind resource is 10% lower than P50 for years 1–5 and is 5% higher for the years 6–15 and
2. 'Extended Operation', where the project operates for 21, 22, 23, 24 and 25 years rather than the 20-year design life.

A spread sheet model was set up which included technical and financial input parameters as shown in Tables 4.24–4.27. These data were used to calculate real and discounted cash flows as shown in Table 4.28. This shows investment cash flows, cash inflows and cash outflows which are used to calculate:

- discounted cumulative cash outflows which give LCCs up to a given year; and
- after-tax cumulative discounted cash flows which give the NPV of the project for each year.

The discount factors for each year for the 7.6% WACC are shown at the bottom of the table.

Table 4.26   Financial parameters for the project.

Debt share	80%
Cost of debt	6%
Equity share	20%
Return on equity	14%
WACC	7.6%

Table 4.27   Wind farm characteristics, revenues and applicable tax rate.

Wind farm characteristics	
Wind farm rate capacity	4.6 MW
Turbine rated capacity	2.3 MW
No. turbines	2
P90 output	12,150 MWh/annum
P50 output	14,350 MWh/annum
P90 capacity factor	30.2%
P50 capacity factor	35.6%
Revenues	
PPA Rate	85 €/MWh
Taxes	
Tax rate	20%

Table 4.28 Cash flows (in €'000s) for the wind farm project.

														Year												
	0	1	2	3	4	5	6	7	8	9	10	11	12	13	14	15	16	17	18	19	20	21	22	23	24	25
Capital expenditure (investment cash flows)																										
Wind farm capital cost	(6665)																									
Income (cash inflows)																										
Electricity sales		1220	1220	1220	1220	1220	1220	1220	1220	1220	1220	1220	1220	1220	1220	1220	1220	1220	1220	1220	1220	1220	1220	1220	1220	1220
Expenditure (cash outflows)																										
Rent		(37)	(37)	(37)	(37)	(37)	(37)	(37)	(37)	(37)	(37)	(37)	(37)	(37)	(37)	(37)	(37)	(37)	(37)	(37)	(37)	(37)	(37)	(37)	(37)	(37)
Rates		(22)	(22)	(22)	(22)	(22)	(22)	(22)	(22)	(22)	(22)	(22)	(22)	(22)	(22)	(22)	(22)	(22)	(22)	(22)	(22)	(22)	(22)	(22)	(22)	(22)
Directors fees		(10)	(10)	(10)	(10)	(10)	(10)	(10)	(10)	(10)	(10)	(10)	(10)	(10)	(10)	(10)	(10)	(10)	(10)	(10)	(10)	(10)	(10)	(10)	(10)	(10)
Management charges		(130)	(130)	(130)	(130)	(130)	(130)	(130)	(130)	(130)	(130)	(130)	(130)	(130)	(130)	(130)	(130)	(130)	(130)	(130)	(130)	(130)	(130)	(130)	(130)	(130)
Insurance		(27)	(27)	(27)	(27)	(27)	(27)	(27)	(27)	(27)	(27)	(27)	(27)	(27)	(27)	(27)	(27)	(27)	(27)	(27)	(27)	(27)	(27)	(27)	(27)	(27)
Accounting fees		(2)	(2)	(2)	(2)	(2)	(2)	(2)	(2)	(2)	(2)	(2)	(2)	(2)	(2)	(2)	(2)	(2)	(2)	(2)	(2)	(2)	(2)	(2)	(2)	(2)
Audit		(3)	(3)	(3)	(3)	(3)	(3)	(3)	(3)	(3)	(3)	(3)	(3)	(3)	(3)	(3)	(3)	(3)	(3)	(3)	(3)	(3)	(3)	(3)	(3)	(3)
General expenses		(3)	(3)	(3)	(3)	(3)	(3)	(3)	(3)	(3)	(3)	(3)	(3)	(3)	(3)	(3)	(3)	(3)	(3)	(3)	(3)	(3)	(3)	(3)	(3)	(3)
Total cash outflows																										
Total cash outflows	(6665)	(234)	(234)	(234)	(234)	(234)	(234)	(234)	(234)	(234)	(234)	(234)	(234)	(234)	(234)	(234)	(234)	(234)	(234)	(234)	(234)	(234)	(234)	(234)	(234)	(234)
Total discounted cash outflows	(6665)	(217)	(202)	(188)	(174)	(162)	(151)	(140)	(130)	(121)	(112)	(104)	(97)	(90)	(84)	(78)	(72)	(67)	(62)	(58)	(54)	(50)	(47)	(43)	(40)	(37)
Cumulative total discounted cash outflows	(6665)	(6882)	(7084)	(7271)	(7446)	(7608)	(7758)	(7898)	(8028)	(8149)	(8261)	(8365)	(8462)	(8553)	(8636)	(8714)	(8787)	(8854)	(8916)	(8974)	(9028)	(9079)	(9125)	(9168)	(9209)	(9246)
Tax effects																										
Profit before depreciation and tax	(6665)	986	986	986	986	986	986	986	986	986	986	986	986	986	986	986	986	986	986	986	986	986	986	986	986	986
Depreciation		(248)	(248)	(248)	(248)	(248)	(248)	(248)	(248)	(248)	(248)	(248)	(248)	(248)	(248)	(248)	(248)	(248)	(248)	(248)	(248)	0	0	0	0	0
Taxable income		738	738	738	738	738	738	738	738	738	738	738	738	738	738	738	738	738	738	738	738	986	986	986	986	986
Tax		(148)	(148)	(148)	(148)	(148)	(148)	(148)	(148)	(148)	(148)	(148)	(148)	(148)	(148)	(148)	(148)	(148)	(148)	(148)	(148)	(197)	(197)	(197)	(197)	(197)
Cash flows																										
Net cash flows after tax	(6665)	839	839	839	839	839	839	839	839	839	839	839	839	839	839	839	839	839	839	839	839	789	789	789	789	789
Discounted cash flows	(6665)	779	724	673	626	581	540	502	467	434	403	375	348	324	301	279	260	241	224	208	194	169	157	146	136	126
Cumulative discounted cash flows	(6665)	(5886)	(5161)	(4488)	(3863)	(3281)	(2741)	(2239)	(1772)	(1339)	(935)	(561)	(213)	111	412	691	951	1192	1416	1625	1819	1988	2146	2292	2428	2554
Discount factors																										
Discount factor	1.000	0.929	0.864	0.803	0.746	0.693	0.644	0.599	0.557	0.517	0.481	0.447	0.415	0.386	0.359	0.333	0.310	0.288	0.268	0.249	0.231	0.215	0.200	0.185	0.172	0.160
Cumulative discount factor		0.929	1.793	2.596	3.342	4.035	4.680	5.278	5.835	6.352	6.833	7.280	7.695	8.081	8.439	8.773	9.082	9.370	9.638	9.886	10.117	10.332	10.532	10.717	10.890	11.050

Table 4.29   NPVs, IRRs and LCOEs for the wind farm project for project lifespans of 20–25 years.

	Lifespan (yr)					
	20	21	22	23	24	25
NPV (€000s)	1819	1988	2146	2292	2428	2554
IRR (%)	11.0	11.2	11.4	11.5	11.7	11.8
LCOE	62.19	61.23	60.38	59.62	58.93	58.31

From Table 4.28, it can be seen that the NPV of the project is €1.819m after year 20. LCC in year 20 is €9.028m, which when divided by cumulative discounted energy output up to year 20 gives an LCOE of €62.19/MWh, lower than the PPA price of 85 €/MWh. Table 4.29 summarises the findings for NPV, IRR and LCOE for the project with lifespans between 20 and 25 years. It can be seen that the NPV increases from €1.819m to €2.554m (40%) over the 5 years. IRR increases from 11.0% to 11.8% while LCOE decreases from 62.19 to 58.31 €/MWh. Each additional year of operation is very valuable, adding 8% to the NPV on average.

Sensitivity analysis results for the 20-year lifespan are shown in Figure 4.5. It can be seen that the NPV of the project is most sensitive to variations in

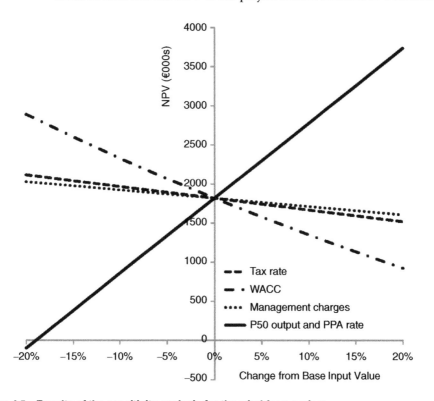

Figure 4.5   Results of the sensitivity analysis for the wind farm project.

revenue resulting from either lower than expected net energy output (P50 output) or lower unit energy prices (PPA rate). This is the reason that the developer has opted to enter into a fixed PPA rather than rely on the sales to the volatile wholesale market. The project NPV is next most sensitive to WACC, the major component of which is debt, which highlights the importance of securing the most competitive debt finance for the project from the banks. The project is relatively less sensitive to tax and management charges.

The model net energy yield for the first 5 years of production is set to 90% of the project P50 energy yield, 105% for years 6–15 and 100% for years 16–20. The simulation results in an NPV of €1.661m (an 8.7% decrease) an IRR of 10.6% (dropping from 11%) and an LCOE of 62.14 €/MWh, marginally down from the original 62.19 €/MWh but still well below the PPA rate of 85 €/MWh. This scenario, therefore, does not represent a significant risk to project viability.

In summary, the positive NPV of €1819m, the IRR of 11% and an LCOE significantly lower than the PPA all indicate an attractive wind farm investment. The greatest risk to the project is caused by a variation in the price paid for the electricity output, which has been mitigated by entering into a 15-year PPA contract with a solid energy supply company. The project is equally exposed to lower-than-expected net energy yields, but this has been mitigated by a 1-year on-site wind resource monitoring and energy yield estimations by a reputable specialist wind consultancy. WACC has a significant but lower impact on NPV than energy yield and prices; this is managed by securing the lowest possible 15-year debt finance backed by a PPA and maintenance contract over the same period. The scenario where energy yields are lower-than-expected in the first 5 years of the project due to variations in wind resource (but equal projections over the long term) does not result in sufficient deterioration in financial parameters to warrant concern. Finally, an additional 5 years of project lifespan will result in a 40% increase in NPV, an important consideration when choosing the specification for the wind turbines.

4.5 Conclusion

In this chapter, we introduced the concept of the time value of money, which represents the risk-weighted opportunity cost of funds invested and explained the superiority of discounted cash flows over undiscounted values. Financial measures which employ these discounted after-tax cash flows are, therefore, better than those which do not. A variety of these measures were presented and discussed with NPV being the truest measure of a project's contribution to increased (or decreased) value for the organisation, society or individual. Other measures employing discounted cash flows have their uses: IRR and MIRR can be compared directly to financial returns on other investment products, LCC can be used to compare costs for projects with identical outputs and LCOE can be used to compare the unit energy supply costs of different technologies. Parameters using undiscounted cash flows such as PP and RoI are inferior.

The financial appraisal of energy projects is based on projected future incremental cash flows and, therefore, financial parameter values are only as good as the cash flow estimations used. The correct choice of the 'base' or 'BAU' case is important when estimating incremental cash flows, and often doing nothing is not a feasible base case scenario. Because the future is unpredictable and because cash flow modelling is never fully accurate, all financial appraisals are subject to significant degrees of uncertainty. Quantifying and understanding this uncertainty is an important step in the financial appraisal process. Sensitivity analysis should always be undertaken while scenario analyses should also be considered to analyse a plausible combination of variations in input parameters. Monte Carlo analysis is normally employed in more complex or academic analyses.

Financial appraisal of a project is a project in itself and should be properly managed. This involves logging assumptions and decisions as well as releasing findings for critical review by competent professionals (either within or outside the organisation). The process normally results in the presentation of findings to the senior management team in an organisation. It is important that they understand the models employed, assumptions made and the project options explored; this information should normally be clearly communicated in written form. Finally, a financial parameter is rarely the only criterion to be considered when deciding whether or not to proceed with a project; many other strategic and operational issues will contribute to the decision.

References

Clark, J., Hindelang, T., & Pritchard, R. (1989) *Capital Budgeting – Planning and Control of Capital Expenditures*, 3rd edn. Prentice-Hall.

Hausman, J. A. (1979) Individual discount rates and the purchase and utilization of energy – using durables, *The Bell Journal of Economics*, Vol. 10, No. 1 (Spring, 1979), pp. 33–54, The RAND Corporation.

Warner, J. and Pleeter, S. (2001) The Personal Discount Rate: Evidence from Military Downsizing Programs. *The American Economic Review*, Vol. 91, No. 1 (Mar., 2001), pp. 33–53. American Economic Association.

5 Multi-Criteria Analysis

5.1 General

With multi-criteria decision analysis (MCDA), the analysis moves away from the concept of the appraisal process being a purely economic one, with the possibility of making some adjustments in order to widen the focus of the study, to the point where the analysis is seen as multi-criteria based, where all attributes, be they monetary or non-monetary, economic or environmental, are assessed on an equal basis. Many complex large-scale energy projects and policies involve attributes that are difficult to both define and measure in money terms. Although some attributes may be quantifiable, it may prove impossible to translate these into monetary values. Others may be intangibles or have qualitative attributes with no attainable quantitative measure of their effects. The appraisal of many public sector engineering development projects involves consideration of such decision criteria. The multi-criteria framework allows such factors to be presented in a comprehensive and consistent format. Although it is a methodology derived from the cost–benefit framework, it has evolved into a decision system that allows all types of attributes to be assessed on an equal footing, possessing many of the characteristics of an effective multi-criteria model.

The overall strategy within multi-criteria decision models involves *decomposition* followed by *aggregation*. The decomposition process divides the problem into a number of smaller problems involving each of the individual criteria. This breaking down of a problem into a number of smaller problems makes it easier for the decision maker to analyse the information coming from diverse origins. The process of aggregation allows all the individual pieces of information to be drawn together to allow a final decision to be made. Within multi-criteria models, the process of aggregation involves

Renewable Energy and Energy Efficiency: Assessment of Projects and Policies, First Edition.
Aidan Duffy, Martin Rogers and Lacour Ayompe.
© 2015 John Wiley & Sons, Ltd. Published 2015 by John Wiley & Sons, Ltd.
Companion Website: www.wiley.com/go/duffy/renewable

either using the information or making certain assumptions concerning the relative importance weightings of the different criteria.

This type of decision model is particularly relevant to the appraisal process for public sector based energy projects or policies where environmental and social criteria must be assessed on a somewhat equal footing to the economic considerations. Use of a process that is based on an ability to include all possible factors rather than one that leads to the exclusion or marginalisation of certain classes of attribute is much more likely to lead to public acceptance of whatever the appraisal reveals.

The techniques examined in this chapter can be broken down into four groups:

1. Simple 'non-compensatory' methods;
2. Simple additive weighting (SAW) method;
3. Analytic hierarchy process (AHP); and
4. Concordance analysis.

The chapter concludes with a case study illustrating the use of a decision model for selecting a wind farm site.

5.2 Simple non-compensatory methods

5.2.1 Introduction

These methods are termed 'simple' because selection of the preferred option does not involve trade off of the disadvantages on one criterion against the advantages on any of the others. There is no process of decomposition followed by aggregation as in the more advanced methods described later in the chapter. Superiority in one criterion cannot be offset by an inferior performance on some other one. It is, therefore, a simpler process but one which leads to less rational outcomes as compared to some of the more complex 'compensating' multi-criteria techniques referred to subsequently. Methods include dominance, satisficing, sequential elimination and attitude-oriented techniques.

With dominance techniques, options are denoted as being members of an exclusive 'non-dominated' set if no other option exists that performs better than it on any of the criteria. Otherwise, the option is deemed to be 'dominated' by another. Satisficing techniques do not actively choose the best options, but merely eliminate those that do not meet certain minimum performances associated with each of the criteria. In the case of sequential elimination techniques, the decision is made by reference to only one criterion, with option choice based on their relative performance on this criterion alone. After the initial evaluation on the first chosen criterion, if more than one option remains, a second criterion can be used to separate them. Attitude-oriented methods take into account the decision maker's attitude towards the environment within which the choice is being made, with this attitude being either pessimistic, in which case selection of an option is based on how badly it performs on its worst scoring criterion, or optimistic, in which case selection is based on how well it performs on its best scoring criterion.

5.2.2 Dominance

An option is dominated if another option exists that performs better than it on one or more of the decision criteria and equals it on the remainder. It is, in effect, a screening or 'sieving' process, where the options being evaluated are reduced to a shortlist by eliminating all those that are dominated.

The sieving process proceeds as follows: compare the first two options; if one is dominated by the other, discard the dominated one. Next, compare the undiscarded option with the third option and again discard the dominated one. Further, introduce the fourth option and so on as before. If the process involves n project options the process requires $n - 1$ steps.

Example 5.1   Dominance Method

Eight alternative residential energy systems are evaluated on the basis of three decision criteria. The options and criteria are briefly described in Tables 5.1 and 5.2, respectively. Table 5.3 details the performance of the eight options on each of the three decision criteria.

Table 5.1   Description of energy system options.

Options	Descriptions
S_1	Conventional energy system
S_2	All electric system
S_3	Energy storage system
S_4	Photovoltaic system
S_5	Wind energy system
S_6	Hybrid energy system
S_7	Gas engine system
S_8	Fuel cell system

Table 5.2   Description of decision criteria.

Criteria	Descriptions	Unit
RC	Running cost of system	€000s
CE	CO_2 emissions of system	Tonnes (t)
EC	Primary energy consumption of system	Gigajoules (GJ)

Table 5.3 Numerical values of various criteria for all options.

Options	Criterion scores		
	RC	CE	EC
S_1	5.22	8.55	219
S_2	3.74	7.35	201
S_3	3.33	8.30	206
S_4	4.09	7.54	198
S_5	3.72	7.31	200
S_6	3.66	7.25	191
S_7	1.83	8.94	216
S_8	1.74	7.29	199

The dominance method is used to derive a preferred non-dominated option or shortlist of non-dominated options. There are two non-dominated options, S_6 and S_8. S_3 and S_7 are dominated by S_8 only, and S_4 is dominated by S_6 only. S_5 is dominated by both S_6 and S_8. S_2 is dominated by three other options, S_5, S_6 and S_8. S_1 is dominated by four other options, S_2, S_5, S_6 and S_8. Using the three chosen criteria, the dominance method derives a shortlist of two options: the hybrid energy system and the fuel cell system.

Figure 5.1 details the relative ranking of the eight options using the dominance method.

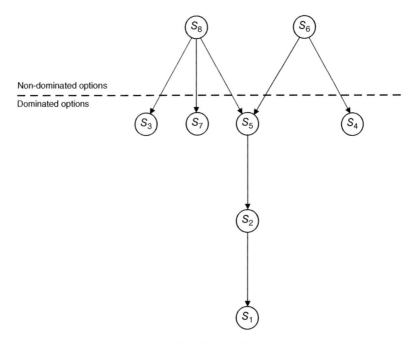

Figure 5.1 Dominance graph for the eight residential energy systems.

5.2.3   Satisficing methods

These methods are not used to identify a chosen best option. They work on the basis of dividing proposals into two categories, acceptable and unacceptable. An option is acceptable if it meets certain minimum standards set by the decision maker. There are two distinct methods within this system – conjunctive and disjunctive.

Conjunctive method

Using this approach, any option that has a criterion valuation less than the standard level will be rejected. For an option to be chosen, it must exceed a minimum value on all criteria. Options are thus easily rejected using this technique; they need to fail on only one criterion. The approach requires the decision maker to generate the minimum cut-off criterion valuations. The setting of these cut-off values is critical to the proper working of the technique. If they are set too high, all options will be deemed unacceptable, whereas if they are set too low too many will remain. Sometimes the standard levels are increased gradually and the group of acceptable options is narrowed down to one choice.

Example 5.2   The Conjunctive Method

A developer wishes to exploit a geothermal resource and devises four alternative scenarios for achieving this objective, ranging from low-level development comprising the development of 2 ha of geothermal greenhouses ('low' option),an intermediate-level development comprising 7 ha of geothermal greenhouses ('intermediate' option), an advanced-level development comprising 7 ha of geothermal greenhouses plus 4 ha of shrimp farming ('advanced' option) and an intense-level development comprising 7 ha of geothermal greenhouses plus 4 ha of shrimp farming plus a hotel complex ('intensive' option).

The four scenarios are evaluated on the basis of five criteria: fuel saved (tons of oil equivalent per year), return on investment (ROI) (yearly earnings over total investment), number of jobs created, environmental impact (measured on a 1–10 scale, with 10 signifying the lowest possible environmental effect) and risk (measured on a 1–10 scale, with 10 signifying the lowest possible risk).

The performances of the four scenarios on each of the five criteria are detailed in Table 5.4.

The decision maker specified the following minimum cut-offs for each of the five criteria:

1. fuel saved ≥1000 toe/year;
2. ROI ≥ 10%;
3. number of jobs created ≥40;
4. environmental index ≥5; and
5. risk index ≥5.

Table 5.4   Performance of the four scenarios on the five decision criteria.

Options	Criterion scores				
	Fuel saved	Return on investment (%)	Number of jobs created	Environmental index	Risk index
Low	200	9	10	9	7
Intermediate	725	10	30	8	6
Advanced	6500	13	50	6	5
Intensive	7250	20	150	4	3

The low option fails to meet the threshold on three criteria (fuel saved, ROI and jobs created), the intermediate and intensive options fail to meet the threshold on two criteria, the former failing on fuel and jobs, with the latter failing on the environmental and risk indices. Only the advanced scenario meets all the thresholds, on the basis that it delivers an acceptable level of economic return and fuel savings while also having acceptable environmental impacts associated with it.

Disjunctive method

An option is chosen in this approach if it exceeds some specified cut-off value on at least one of the criteria. In effect, an option is deemed acceptable if it performs to a high level on one of the criteria. Those that do not perform exceptionally well on any of the decision criteria are not selected under this process.

Example 5.3   The Disjunctive Method

Taking the same problem as Example 5.2, the decision makers specify the following desirable scores for each of the five criteria:

1. fuel saved $\geq$10,000 toe/year;
2. ROI $\geq$ 20%
3. number of jobs created $\geq$ 100;
4. environmental index $= 10$; and
5. risk index $= 10$.

One can see that, using the disjunctive method, the 'intensive' scenario is the only one delivering an exceptional result on any of the five criteria, delivering a return of 20% and creating 150 jobs.

5.2.4   Sequential elimination methods

There are two approaches that eliminate options in a sequential way: the lexicographic and elimination by aspects techniques. Both use one criterion at a time to evaluate the options, excluding all except the best performing one. If a tie exists between options, it is resolved by examining their relative performance on another criterion.

Lexicographic method

In this method, the options are examined initially on the criterion deemed to be the most important. It is used mostly in cases where one criterion predominates. If one option has a better score on the chosen criterion as compared to all other options, it is chosen and the process ends. If, however, a number of options are tied on the most important criterion, this subset of tied options is then compared on the next most important criterion. The process continues in a sequential manner until a single option is identified or all criteria have been gone through and complete separation proves impossible.

For example, take the situation where a number of project options are being evaluated and cost is seen as the most important criterion. Only where the cost for two options is equal does some other criterion come into play to separate them. This method can be altered slightly by stating that differences in scores between options on a given criterion must be greater than a certain threshold before they are deemed significant. This variation is called a 'lexicographic semi-order'.

Example 5.4   The Lexicographic Method

Six policy options are proposed for developing renewable energy sources within a region, wind, biomass, photovoltaic (PV), a combination of wind–biomass, a combination of wind–PV and a combination of wind–biomass–PV.

The five criteria used to evaluate the six options are detailed in Table 5.5.

Table 5.5   Criterion descriptions and qualitative scale used.

Criteria	Descriptions	Scale
Employment	Increases in employment during construction and operation	1–5
Land used	Area of land used for each renewable energy system option	1–5
Visual	Visual disturbance by facility on the landscape	1–5
Emissions	CO_2, SO_2 and NO_x mitigation potential	1–5
Efficiency	Useful energy obtained from energy source	1–5

Table 5.6 details the performance of the six local policy options on each of the five criteria.

Table 5.6 Performance of the six options on the five decision criteria.

Options	Employment	Land	Visual	Emissions	Efficiency
Wind	3	2	1	5	5
Biomass	2	3	3	1	1
PV	1	1	2	4	3
Wind–bio	3	3	2	4	2
Wind–PV	2	3	1	4	3
Wind–bio–PV	3	3	1	1	2

The following order of importance has been allocated to the five criteria:

1. Land used;
2. Employment;
3. Emissions;
4. Visual; and
5. Efficiency.

Basic lexicographic method

On the most important criterion (land used), four options are tied with a score of 3, biomass, wind–bio, wind–PV and wind–Bio–PV. On the second most important criterion (employment), two of these four options are tied on a score of 3, wind–bio and wind–bio–PV. On the third most important criterion (emissions), wind–bio with a score of 4 is preferred to wind–bio–PV with a score of 1, and is thus the chosen option using this technique.

Lexicographic semi-order

The threshold on each criterion is a score difference of more than 1.

On this basis, on the most important criterion (land used), five options are now tied with a score of 2 or 3, wind, biomass, wind–bio, wind–PV and wind–bio–PV. On the second most important criterion (employment), all five score either 2 or 3. On the third most important criterion (emissions), three options, wind, wind–bio and wind–PV have a score of 4 or 5. On the fourth most important option (visual), all three remaining options have a score of 1 or 2. On the fifth most important criterion (efficiency), wind with a score of 5 is preferred to wind–bio and wind–PV with scores of 2 and 3, respectively, and is thus the chosen option using this technique.

5.2.5 Attitude-oriented methods

The following two non-compensatory models take into account the decision maker's attitude towards the environment within which the choice is being made. In the case of the Maximin technique, the decision maker has a pessimistic attitude; therefore, the worst performing criterion for each option is identified, and the option that scores best on its 'worst criterion' is chosen. With Maximax, an optimistic attitude prevails. Here, the option scoring best on its 'best criterion' is chosen.

Maximin

Any renewable energy policy or project is only as strong as its weakest link. There may be circumstances where the strength of any given proposal will be gauged by how well it performs on its weakest decision criterion.

Where a decision maker does not have any prior information regarding which criterion will have the greatest influence on overall performances, a pessimistic attitude should be adopted where the option whose worst score is better than the worst performances of the others is chosen. In other words, the 'best of the worst' is being selected.

An option is thus represented by its single worst criterion performance. All other criterion scores are disregarded. As we may be comparing scores from different criteria, the technique only works where all criteria are measured on a common scale that can be quantitative or qualitative.

The selection process involves two steps:

1. Determine the worst criterion score for each option and
2. Choose the option with the best score on its worst criterion.

Example 5.5   Maximin

Taking Example 5.2, but reconfiguring all criterion scores on a scale 1–10, the resulting performance matrix for the four alternative geothermal resource exploitation scenarios are as detailed within Table 5.7.

Table 5.7   Performance of the four scenarios on the five decision criteria (all criteria scored 1–10).

Options	Criterion scores				
	Fuel saved	Return on investment	Number of jobs created	Environmental index	Risk index
Low	2	4	1	9	7
Intermediate	3	5	3	8	6
Advanced	8	7	5	6	5
Intensive	10	10	9	4	3

Use of this decision model is appropriate if the decision maker is of the opinion that, no matter how good an option's overall performance is, if it performs badly on even one criterion, that option may attract widespread public opposition. An option that performs consistently on all criteria is thus preferred to one that performs extremely well and quite poorly on others.

Examining the performances of the four scenarios in Table 5.7, one can see that the 'low' scenario has a lowest score of 1, the 'intermediate' a lowest score of 3, the 'advanced' a lowest score of 5 and the 'intensive' a lowest score of 3. The highest 'worst' score is attributed to the 'advanced' scenario which scores a minimum of 5 on all criteria. This option is thus chosen on the basis that it is likely to be the most widely acceptable due to its consistent scoring.

Maximax

In contrast to Maximin, this technique chooses an option based on its best criterion score rather than its worst. The method is useful in situations where options can be chosen for specific functions based on its performance on any one of the particular criteria in question.

Again, within this method, only one criterion score represents each option under consideration. All other criterion performances are disregarded. Also, it too requires that all criteria be scored on a common scale.

Maximax has two operating procedures also:

1. Determine the best criterion score for each option and
2. Choose the option with the best score on its best criterion.

Example 5.6   Maximax

Again taking the performance matrix for the four alternative geothermal resource exploitation scenarios are as detailed within Table 5.7 within Example 5.5, if the decision maker decides that the preferred option should be chosen based on an outstanding performance on one of the chosen criteria, then the 'advanced' scenario is preferred on the basis that it scores a maximum 10 on two of the five criteria.

5.3   Simple additive weighting method

5.3.1   Basic simple additive weighting method

The SAW method is a classical 'compensatory' model (Keeney and Raiffa, 1976). Compensation allows an option's performance on one criterion to be traded off against its performance on another. To allow compensation to take place, the performances of a given option on the different criteria must be put onto one common scale of measurement. The individual criterion scores for each option can then be manipulated mathematically in order to compute an index of overall performance for each one. These indices enable the options to be ranked relative to each other. The additive weighting method is the simplest form of a more general decision model based on multi-attribute utility theory (MAUT). MAUT, or utility theory, is widely used to solve project appraisal problems, not only in the engineering sector but also in the financial and actuarial spheres. It involves devising a function that will express the utility of a project option in terms of a number of agreed relevant decision criteria (Keeney and Raiffa, 1976). It is not considered further within this book.

The SAW method is one of the prominent and most widely used multi-criteria methods. In effect, an overall score for an option is obtained by adding the contributions from each of the chosen criteria. As two criteria with different measurement scales cannot be added, a common numerical scaling system is required to allow the addition of the different criterion performances for each option. The total overall score for each option is estimated by multiplying the option score on each criterion by the importance weighting of that criterion and then summing these products over all the criteria in question.

The overall weighted score, V_i, for a project option i, using this method can be written as follows:

$$V_i = \sum_{j=1}^{j=n} w_j r_{ij} \qquad (5.1)$$

where
w_j = weight for criterion j
r_{ij} = score for option i on criterion j.

The option with the largest value of V_i is selected.

The additive weighting method thus converts the multi-criterion problem to a single-dimension problem. Again, in the vast majority of complex energy project and policy appraisals, the different criterion types being assessed within the process, because of their diverse origins (economic, environmental, technical) are expressed in various units of measurement. These must then be converted to a common scale before the additive model can be utilised. A common procedure employed is to convert all criterion scores to a normalised linear scale going from 0 (worst) to 1 (best), though this is often magnified to a 10-point or a 100-point scale.

A weighting for the cost criterion must be derived, which reflects its importance relative to the others deemed relevant to the study. Again, a common procedure employed is to convert all weighting scores to a normalised linear scale going from 0 (least important) to 1 (most important). The weighting scores are then in a form that can be included within the additive weighting model, with a single overall score being obtainable for each option examined.

Example 5.7 Baseline Simple Additive Weighting Method

Taking Example 5.4, but reconfiguring all criterion scores on a scale 1–10, the resulting performance matrix for the six alternative energy systems are as detailed in Table 5.8.

Table 5.8 Performance of the six options on the five decision criteria (unweighted scores).

Options	Employment	Land	Visual	Emissions	Efficiency
Wind	6	4	2	10	10
Biomass	4	6	6	2	2
PV	2	2	4	8	6
Wind–bio	6	6	4	8	4
Wind–PV	4	6	2	8	6
Wind–bio–PV	6	6	2	2	4

The normalised weightings for the five criteria are detailed in Table 5.9.

Table 5.9 Criterion weightings.

Criterion	Weight
Land used	0.30
Employment	0.25
Emissions	0.20
Visual	0.15
Efficiency	0.10

The final step involves the multiplication of the ratings by the relevant criterion weights. This calculation is shown in Table 5.10. It can be seen that two options, wind and wind–bio, perform well, with wind–PV in third place. All these options are placed well ahead of the other three.

Table 5.10 Overall scores for the three options.

Criterion	Weight	Wind	Biomass	PV	Wind–bio	Wind–PV	Wind–bio–PV
Land used	0.30	4	6	2	6	6	6
Employment	0.25	6	4	2	6	4	6
Emissions	0.20	10	2	8	8	8	2
Visual	0.15	2	6	4	4	2	2
Efficiency	0.10	10	2	6	4	6	4
Score		6.0	**4.3**	3.9	5.9	5.3	4.4

5.3.2   Sensitivity analysis of baseline SAW results

The data that is input into a decision model is seldom known with complete certainty. A sensitivity analysis allows the decision maker to gauge the effect that incremental changes in criterion weights and valuations will have on the final result. The importance weightings are of particular interest within a sensitivity analysis as their valuation is the result of subjective judgements by experts who, because of their diverse backgrounds, may disagree about their correct value.

With criterion scores, sensitivity testing arises from errors that may arise from the actual estimation of the valuations themselves. They may be based on incomplete data or, in the case of qualitative assessments, may be derived from subjective judgments by specialists from the relevant field. Their exact value may therefore not be known with complete confidence

The process can be arduous and time consuming, involving a large number of iterations, as different criterion and weighting scores are varied incrementally and the effect of these changes on the original ranking of options is gauged. Incremental changes in criterion valuations possessing a high weighting score are more likely to alter the baseline ranking than changes in the scores of less significant criteria. Particular attention may also be paid to criterion scores that involve a high degree of uncertainty and subjectivity. A competent decision maker should identify such criteria and analyse the effect of their variation on the overall ranking.

Example 5.8   Sensitivity Analysis within Simple Additive Weighting Method

Taking the baseline example within Example 5.7, six sensitivity tests are carried out on the baseline weighting scores, where the three most important criteria are subjected to a ±50% variation. The effect of these variations on the overall rankings is assessed.

The analysis is therefore broken down into the following six sensitivity tests:

Test 1 – 'Land used' criterion +50%, all other weights unchanged
Test 2 – 'Employment' criterion +50%, all other weights unchanged
Test 3 – 'Emissions' criterion +50%, all other weights unchanged
Test 4 – 'Land used' criterion −50%, all other weights unchanged
Test 5 – 'Employment' criterion −50%, all other weights unchanged
Test 6 – 'Emissions' criterion −50%, all other weights unchanged.

Table 5.11 details the weighting system derived from these adjustments, while Table 5.12 shows the overall scores and rankings for each of the options on each of the six sensitivity tests.

Table 5.11   Criterion weightings within the six sensitivity tests.

Criterion	Normalised criterion weightings					
	Test 1	Test 2	Test 3	Test 4	Test 5	Test 6
Land used	0.39	0.27	0.27	0.18	0.34	0.33
Employment	0.22	0.33	0.23	0.29	0.14	0.28
Emissions	0.17	0.18	0.27	0.24	0.23	0.11
Visual	0.13	0.13	0.14	0.18	0.17	0.17
Efficiency	0.09	0.09	0.09	0.12	0.11	0.11

Table 5.12   Results of sensitivity tests.

Option	Sensitivity tests						Avg. rank	Baseline ranking
	Test 1	Test 2	Test 3	Test 4	Test 5	Test 6		
Wind	5.7 (2nd)	6.0 (1st)	6.4 (1st)	6.4 (1st)	6.0 (1st)	5.6 (2nd)	1.33	1st
Biomass	4.5 (5th)	4.3 (5th)	4.1 (6th)	4.0 (6th)	4.3 (5th)	4.6 (5th)	5.33	5th
PV	3.7 (6th)	3.7 (6th)	4.3 (4th)	4.2 (4th)	4.2 (6th)	3.4 (6th)	5.33	6th
Wind-Bio	5.9 (1st)	5.9 (2nd)	6.1 (2nd)	5.9 (2nd)	5.9 (2nd)	5.7 (1st)	1.67	2nd
Wind–PV	5.4 (3rd)	5.2 (3rd)	5.5 (3rd)	5.2 (3rd)	5.5 (3rd)	5.0 (3rd)	3.00	3rd
Wind-B-PV	4.6 (4th)	4.6 (4th)	4.2 (5th)	4.1 (5th)	4.2 (4th)	4.7 (4th)	4.33	4th

The sensitivity analysis confirms the closeness of the top two options: wind and wind–bio. Wind is ranked first on four of the six tests, with wind–bio placed first in the other two. The other four options are consistent in their rankings relative to each other, all performing below the top two.

5.3.3   Assigning weights to the decision criteria

The assignment of weights to the constituent criteria is central to the working of the additive model. Their purpose is to express the importance of each chosen decision criterion relative to all others. A number of basic techniques exist for calculating criterion weightings within the additive model. All require the input of judgements from the decision makers.

The following methods are outlined below:

- Presumption of equal weights;
- Ranking system for obtaining weights;
- Ratio system for obtaining weights; and
- Resistance-to-change grid.

Presumption of equal weights

If the decision makers are not in a position to assign weights to the criteria, or are unwilling to do so, the process can proceed initially on the basis that all criteria are treated as being of equal importance, with each attribute given an equal weight. In these circumstances, it is particularly important to carry out an extensive sensitivity analysis to gauge the effect of varying the weightings away from their equal valuations on the baseline performances of the project options.

Ranking system for obtaining weights

This procedure involves the decision maker initially ranking the criteria in order of importance. Each criterion is then assigned a score based on its rank, with the one ranked first assigned the score '1', the one ranked second assigned the score '2' and so on. In the case of criteria tying for the same rank, an average score is assigned to them. For example, take the case where one criterion is ranked first and assigned the score '1'. Three criteria are ranked equal second. Their score is obtained by averaging the three scores '2', '3' and '4'. Each of the tied criteria will thus be given the score '3'. The criterion ranked below these three in fifth place will be assigned the next highest score of '5'. The least important criterion will thus end up with the score n, where n is the number of criteria.

The normalised importance weight for each criterion can be calculated using the following formula:

$$w_i = \frac{n - r_i + 1}{\sum_{i=1}^{n}(n - r_i + 1)}$$ (5.2)

where
w_i = normalised weighting for the ith criterion
r_i = ranking score for the ith criterion
n = number of decision criteria.

Example 5.9   Use of Ranking System to Derive Criterion Weights

A number of residential energy systems are being compared on the basis of the following four criteria:

- Investment cost (IC);
- Running cost (RC);
- CO_2 emissions (CE); and
- Primary energy consumption (EC).

The decision makers have agreed a ranking of these criteria from most to least important as detailed within Table 5.13.

Table 5.13 Computation of normalised weights using ranking method.

Criteria	Rank position	Rank score (r_1)	$n-r_1+1$	w_1
EC	First	1	4	0.40
CE,RC	Second	$(2+3) \div 2$	2.5(×2)	0.25
IC	Third	4	1	0.10
			$\Sigma = 10$	

As can be seen from Table 5.13, the primary energy consumption criterion is deemed the most important and is ranked first and assigned the rank score '1'. Running cost and CO_2 emissions are ranked equal second and given the average of the scores '2.5'. Investment cost is ranked fourth and last, and is assigned the score '4'.

Table 5.13 then uses Equation 5.2 to determine the normalised weight for each criterion using this method.

Ratio system for obtaining weights

This technique is similar in nature to the ranking system outlined immediately above. The computation starts by assigning a score of '1' to the criterion or criteria adjudged by the decision makers to be least important. They are then requested to give the other criteria scores greater than '1', with the number of times the assigned score being greater than unity reflecting the importance of that criterion relative to the least important one or ones. Sometimes the decision makers will place a ceiling on the multiple separating the least and most important criteria, for example, deciding that the most important criterion will be three times the weighting of the least important one. These weights can then be normalised so that they sum to unity by using the following equation:

$$w_i = \frac{z_i}{\sum_{i=1}^{n} z_i} \tag{5.3}$$

where
w_i = normalised weighting for the ith criterion
z_i = weight score assigned to ith criterion
n = number of decision criteria.

Example 5.10   Use of Ratio System to Derive Criterion Weights

Analysing the same problem as outlined in Example 5.9, it is decided that the most important criterion (EC) will be separated from the least important (IC) by a factor of three. Thus, EC is assigned '3' and IC '1'. RC and CE are both assigned '1.5'.

Table 5.14 outlines the computation of the normalised weights for each criterion from these raw scores. The final values are quite similar to those obtained from the ranking method.

Table 5.14   Computation of normalised weights using ranking method.

Criteria	Rank position	Ratio score (z_1)	w_1
EC	First	3	0.43
CE,RC	Second	1.5	0.21
IC	Third	1	0.14
		$\Sigma = 7$	

The resistance-to-change grid

This technique, devised by Rogers and Bruen (1998), has a firm methodological basis which ensures that the weights obtained reflect the decision makers actual preferences in terms of what they deem to be important. The previous two techniques, where decision makers spontaneously award weight scores to each of the criteria, do not necessarily relate the actual scores given for each criterion directly back to the basic concept of its importance weighting. The resistance-to-change methodology results in decision makers automatically placing the criteria into a hierarchy of relative importance. This hierarchy can then be used to compute directly the weighting for each.

The key to using this method lies in deriving two terms for each criterion, one expressing its most desirable aspect and the other expressing its least desirable aspect. This is termed expressing the criterion in 'bipolar form'. The four criteria from the last two examples can be expressed in bipolar form as follows:

EC – Low energy consumption/high energy consumption;
CE – Low CO_2 emissions/high CO_2 emissions;
RC – Low running costs/high running costs;
IC – Low investment costs/high investment costs.

The left-hand side of each of the aforementioned bipolar expressions represents the criterion at its most desirable, while the right-hand side represents it at its least desirable. For an economic criterion, the preferred side is assumed to be the one minimising cost. For an environmental criterion, the preferred side is assumed to be the one minimising environmental impact.

In order to derive a pairwise comparison matrix of the type illustrated in Table 5.15 for each pair of criteria, the decision maker is required to examine the bipolar expressions for both. He is then asked the question that, if he had to change one of these criteria from its desirable side to its undesirable side, which one would he be least willing to change. For each comparison, the criterion resisting change is given the score '1', while the criterion which the

Table 5.15   Sample of 'resistance-to-change' grid.

	EC	CE	RC	IC	Row sum	Norm weight
EC		1	1	1	3	0.50
CE	0		0.5	1	1.5	0.25
RC	0	0.5		0.5	1.0	0.17
IC	0	0	0.5		0.5	0.08
					$\Sigma = 6$	

decision maker is willing to change to its undesirable side is given the score '0'. If the decision maker has an equal resistance to changing either criterion, each is given the score '0.5'. When a given criterion has been compared for resistance to change with all others, its scores are totalled to give a final weighting. All weightings obtained are then summed and each is divided by this figure to attain normalisation.

The theoretical basis for the grid originated from Hinkle (1965), who found that the more important a criterion is, the more likely it will be that the decision maker resists changing it to its more undesirable state.

The final weights obtained using the 'resistance-to-change' grid are detailed in Table 5.15. For example, the decision maker, when asked to choose between low energy consumption (EC) and high running costs and high energy consumption and low running costs (RC), is willing to tolerate high running costs as long as energy consumption is low. EC is thus given a score of 1 against RC, with RC given a score of 0 against EC.

Thus, 1, 0 or 0.5 is inserted within in cell in the decision matrix.

5.4   Analytic hierarchy process

5.4.1   Introduction

The AHP works by establishing hierarchies within the problem under examination. It breaks down the decision into a number of discrete elements within each level of the hierarchy and then uses a pairwise comparison methodology to establish priorities both within and between the hierarchies put in place by the decision maker. On each criterion in question, the relative merit of each project option is determined from a pairwise analysis of the scores for all combinations of options. A similar pairwise analysis is also used to determine the relative importance of the criteria. A combination of these two processes yields a relative ranking of all options (Saaty, 1977, 1980). To understand more of the AHP method, we must define the nature of the hierarchies within it, and how the priorities within and between the hierarchies are established.

5.4.2   Hierarchies

A hierarchy enables the decision problem to be broken down into individual elements whose relationships with each other can then be analysed. Stated more formally by Saaty, it is 'an abstraction of the *structure* of a system to study the *functional* interactions of its components and their impacts on the entire system'. The 'structure' and 'function' of a system cannot be separated. The former is the arrangement of its parts, and the latter is the function or duty that the components are meant to serve.

Thus, a hierarchical system is based on the assumption that the entities identified as relevant to the decision can be grouped into disjoint sets, with each set directly affecting the one above it. This is the 'decomposition' referred to in the opening definition. This hierarchical structure illustrates how results at the higher levels are directly influenced by those at the lower levels.

Hierarchies have many advantages. They can be used to describe how changes in priority at upper levels affect priorities of elements in lower levels. They provide detailed information on both the structure and function of the system. They are stable and flexible, accurately mirroring reality, because natural systems are assembled hierarchically.

5.4.3   Establishing priorities within hierarchies

A hierarchy, by itself, is not a very powerful tool for decision making unless the extent to which the various elements on one level influence those on the next higher level can be determined. In this way, the relative strengths of the impacts of the elements at the lowest level on the overall objectives can be computed. This is the 'synthesis' referred to in the opening definition.

The strengths of priorities of elements in one level relative to the next are determined as follows. Given the elements of one level, and one element on the next higher level, the elements of the lower level are compared pairwise in their strength of influence on X. The numbers reflecting these comparisons are inserted in a matrix, and the eigenvector with the largest eigenvalue is found. The eigenvector itself provides the priority ordering, and the eigenvalue is a measure of the consistency of the judgement.

For example, let us examine four elements A, B, C and D within one hierarchy level. These could denote project options that are being pairwise compared on a given criterion, or they could represent criteria whose relative importance is being assessed. Each pair – AB, AC, AD, BC, BD and CD – is directly compared with respect to their influence on X, and the results placed within a matrix format are summarized in Table 5.16.

Take a pairwise comparison of A and B for demonstration purposes (see Table 5.17).

The appropriate score is placed in the position 'ab' where row A meets column B.

Table 5.16   Example of pairwise comparison matrix.

	A	B	C	D
A	1	ab	ac	ad
B	1/ab	1	bc	bd
C	1/ac	1/bc	1	cd
D	1/ad	1/bd	1/cd	1

Table 5.17   Example of pairwise comparison matrix.

Pairwise evaluation	Score
A and B are equally important	ab = 1
A is weakly more important than B	ab = 3
B is weakly more important than A	ab = 1/3
A is strongly more important than B	ab = 5
B is strongly more important than A	ab = 1/5
A is very strongly more important than B	ab = 7
B is very strongly more important than A	ab = 1/7
A is absolutely more important than B	ab = 9
B is absolutely more important than A	ab = 1/9

As an element is always equally important relative to itself (reflexivity), the main diagonal of the matrix always consists of 1's. Moreover, the appropriate reciprocals 1/3, 1/5, 1/7 or 1/9 should be inserted where column A meets row B, that is, in position (B, A). In addition, the scores 2, 4, 6 and 8 and their reciprocals are used to facilitate compromise between slightly differing judgements.

The requirement for a nine point scale is based on Saaty's belief that, within the framework of a simultaneous comparison, one does not need more than nine scale points to distinguish between stimuli (Saaty, 1977). Results from psychological studies (Miller, 1956) have shown that a scale of about seven points was sufficiently discerning. Saaty (1977) noted that the ability to make qualitative decisions was well represented by five attributes:

- Equal;
- Weak;
- Strong;
- Very strong; and
- Absolute.

If we make comparisons between adjacent attributes, when greater precision is needed, the scale requires, in total, nine values – thus a nine-point scale.

5.4.4   Establishing and calculating priorities

The point has now been reached where the comparison matrix is in place and the information contained within it is required to establish priorities between the different elements being examined. In order to derive these, the judgements made within the pairwise matrix must be pulled together in order that a single number for each element, indicating its priority relative to the others, can be determined.

Take the following example. A government wishes to decide which of three energy efficiency policy initiatives it should implement. The decision is being made on the basis of one criterion – total cost. Using Saaty's nine-point scale, the policy maker compares the three initiatives, A1, A2 and A3 pairwise:

A1 – Energy management in major buildings
A2 – Improvement in manufacturing processes
A3 – Implementation of night rate campaign

The matrix lists the options within the left column and along the top row (see Table 5.18). The score along the diagonal positions is always '1', as an option is at all times as equally desirable as itself. The judgements above the diagonal are then input. Once these are decided, all judgements below the diagonal become the reciprocal of those above. If the decision involved n options, then the number of judgements needed to fill the entries is as follows:

$$\text{Number of entries} = (n^2 - n) \div 2$$

In this case, the number of elements is 3; therefore, the number of entries required is

$$(3^2 - 3) \div 2 = 3$$

The three required entries in this case are indicated in bold. All other entries are either '1' along the diagonal positions or are reciprocals of the original three above the diagonal.

The government assesses that the cost of option A1 is marginally better than option A2 and assigns it the score '2' on the Saaty scale of 9 in this pairwise comparison (between 'equal' and 'weakly preferable'). Option A1 is assessed

Table 5.18   Performance matrix for energy efficiency initiatives based on cost.

Cost	A1	A2	A3
A1	1	2	4
A2	1/2	1	2
A3	1/4	1/2	1

moderately better than option A3 in terms of cost and is assigned the score '4'. Option A2 is assessed as being marginally better than option A3 in terms of cost and is thus also given the score '2'. These are all the judgements we need to make. We place '1' along the diagonal positions. The other values are reciprocals of our original judgements. The performance of A2 against A1 is the reciprocal of the score for A1 versus A2. A2 is adjudged slightly less preferable in terms of cost as compared to A1 and is assigned the reciprocal of '2', that is, '1/2'. For the same reason A3 is assigned the score '1/4' against A1 and '1/2' against A2.

5.4.5 Deriving priorities using an approximation method

The next step involves translating the scores that have been assigned to each pair of elements within the matrix into a single score for each element indicating its relative performance. For the particular case illustrated in Table 5.18, the aim is to establish the relative priorities of the three energy efficiency options on the basis of cost.

Let us first illustrate a procedure for obtaining an approximate estimate for these priorities. (A more widely applicable and comprehensive technique is outlined further.)

Initially, add the scores within each column and then take each of these column totals and divide it into each entry within its column. This gives a set of normalised scores along each column length, that is, their sum equals one (see Tables 5.19 and 5.20).

In order to get the final scores, we obtain the average entry score for each row of the normalised matrix in Table 5.21.

This simple procedure is sufficient for calculating the priorities of the various elements for the special case where the matrix is consistent. For the

Table 5.19   Judgements with column totals.

Cost	A1	A2	A3
A1	1	2	4
A2	1/2	1	2
A3	1/4	1/2	1
Column total	7/4	7/2	7

Table 5.20   Normalised judgement matrix.

Cost	S1	S2	S3
S1	4/7	4/7	4/7
S2	2/7	2/7	2/7
S3	1/7	1/7	1/7

Table 5.21   Result for each option.

Element	Calculation	Result
A1	$(4/7 + 4/7 + 4/7) \div 3$	0.571
A2	$(2/7 + 2/7 + 2/7) \div 3$	0.286
A3	$(1/7 + 1/7 + 1/7) \div 3$	0.143
	$\Sigma =$	1.0

example shown earlier, a quick calculation indicates the consistency of the comparison matrix:

- A1 versus A2 yields the result '2' (A1 = 2 * A2)
- A1 versus A3 yields the result '4' (A1 = 4 * A3)

Therefore, for consistency,

- A2 versus A3 should yield the result 2 * A2 = 4 * A3 ⇒ A2 = (4/2 * A3) = 2 * A3

We see from the comparison matrix that this is, in fact, the case, because A2 versus A3 gives the answer '2'. The matrix is, therefore, consistent.

In the case where the matrix is not consistent, however, the exact method of determining priorities must be used.

Let us adjust slightly the priorities within the matrix in Table 5.18.

This matrix is not consistent, because although A1 = 2 * A2 and A1 = 5 * S3, A2 ≠ 5/2 * A3. The matrix, in fact judges A2 to be 2 * A3. If we use the approximate method in this case, we will derive the following priorities.

One can see immediately that, unlike the last example, the columns are not identical. The matrix is thus confirmed as being inconsistent and if one obtains the average value for each row as shown in Table 5.24 one cannot be certain that an accurate valuation of the priorities has been obtained:

To obtain, with certainty, accurate estimates of the priorities of the elements in question the method outlined immediately below must be used.

5.4.6   Deriving exact priorities using the iterative Eigenvector method

In order to obtain the exact priorities of the elements where the judgement matrix is not consistent, the principal eigenvector of the matrix must be obtained. The method entails obtaining an initial set of priorities using the approximate method illustrated above. These weights are then input into an n value column vector, where n equals the number of elements under examination. This column vector is then multiplied by the $n \times n$ matrix of judgements. The resulting column vector is then normalised to sum to unity, yielding an updated set of n priorities. The process is then repeated. The updated normalised vector is multiplied by the $n \times n$ matrix of judgements and the resulting column vector is again normalised and compared with the first set of values. If the difference between the two sets of priorities is within a prescribed decimal accuracy, the process is halted and the last set of computed priorities yields what is termed the principal eigenvector of the judgement matrix. If the difference between the two successively derived sets of priorities is not within the prescribed limits, the process is repeated until the required accuracy is obtained.

If the judgement matrix is consistent, the values from the exact method are identical to those from the approximate method. The greater the level of consistency throughout the matrix, the greater the divergence between the results from the two methods will be.

Take the judgement matrix in Table 5.22 that has been shown in Table 5.23 to be inconsistent. The priorities derived by the approximate method, given in Table 5.24 are therefore incorrect. These values can, however, be used within the initial iteration of the exact eigenvector method. These three priorities (0.59, 0.28, 0.13) are placed within a column vector and multiplied by the judgement matrix in Table 5.25.

The resulting column vector is now normalised to give a revised set of priorities for the three options A1, A2 and A3 (Table 5.26).

These revised set of priorities are now multiplied by the basic judgement matrix, and the procedure is repeated in Table 5.27.

The column vector is again normalised (Table 5.28).

It can be seen that the difference between the priorities from Step 1 and Step 2 is minimal. The iterative procedure can therefore be terminated and the values from Step 2 taken as the calculated priorities from the exact method.

Table 5.22 Example of an inconsistent judgement matrix.

Cost	A1	A2	A3
A1	1	2	5
A2	1/2	1	2
A3	1/5	1/2	1

Table 5.23 Normalised matrix.

Cost	A1	A2	A3		A1	A2	A3
A1	1	2	5		0.59	0.57	0.63
A2	1/2	1	2	=	0.29	0.29	0.25
A3	1/5	1/2	1		0.12	0.14	0.12
Column total					$\Sigma = 1$	$\Sigma = 1$	$\Sigma = 1$

Table 5.24 Priorities derived using approximate method.

Option	Calculation	Result
A1	$(0.59 + 0.57 + 0.63) \div 3$	0.5949
A2	$(0.29 + 0.29 + 0.25) \div 3$	0.2766
A3	$(0.12 + 0.14 + 0.12) \div 3$	0.1285
	$\Sigma =$	1.0000

Table 5.25 Column vector computation – step 1.

1	2	5		0.59		1.7906
1/2	1	2	$\times$	0.28	$=$	0.8311
1/5	1/2	1		0.13		0.3858

Table 5.26 Normalisation – step 1.

	1.7906		0.5954
	0.8311	Normalised $\rightarrow$	0.2763
	0.3858		0.1283
Σ	3.0075		1

Table 5.27 Column vector computation – step 2.

1	2	5		0.5954		1.7894
1/2	1	2	$\times$	0.2783	$=$	0.8306
1/5	1/2	1		0.1283		0.3855

Table 5.28 Normalisation – step 2.

	1.7894		0.5954
	0.8306	Normalised $\rightarrow$	0.2764
	0.3855		0.1283
Σ	3.0055		1

The derived priorities can thus be taken as 0.5954, 0.2764 and 0.1283 for options A1, A2 and A3, respectively. These are seen as different from those derived from the approximation method given in Table 5.24.

Note: If the exact method is used to derive the priorities for a totally consistent matrix of judgements, the values from the approximate method are immediately replicated within one step of the iterative procedure, and would continue to do so no matter how many times the procedure is repeated.

Example 5.11   Ranking of Six Energy Systems Using AHP

This example uses AHP to evaluate six alternative residential energy systems on the basis of three decision criteria (Tables 5.29 and 5.30). The options and criteria are briefly described in Tables 5.1 and 5.2, respectively. Table 5.3 details the performance of the eight options on each of the three decision criteria. (This is an abbreviated version of Example 5.1. Only six of the eight options are used here.)

Table 5.29   Description of energy system options.

Options	Descriptions
A_1	Energy storage system
A_2	Photovoltaic system
A_3	Wind energy system
A_4	Hybrid energy system
A_5	Gas engine system
A_6	Fuel cell system

Table 5.30   Description of decision criteria.

Criteria	Descriptions	Unit
RC	Running cost of system	€000s
CE	CO_2 emissions of system	t
EC	Primary energy consumption of system	GJ

The problem involves using AHP at two levels as follows:

- Estimating the relative importance weights of the decision criteria and
- Determining the relative performance of each option on each of the decision criteria.

Judgement matrices

Estimating the relative weights for decision criteria RC, CE and EC.

Table 5.31 is the pairwise matrix of judgements from the decision makers regarding the relative importance weightings of the three decision criteria.

One can see from the normalised matrix in Table 5.32 that not all the columns are identical; therefore the judgements are not fully consistent.

Table 5.31   Judgement matrix for decision criteria.

Criterion	RC	CE	EC
RC	1	1/2	1/6
CE	2	1	1/2
EC	6	2	1

Table 5.32   Normalised matrix.

Criterion	RC	CE	EC		RC	CE	EC
RC	1	1/2	1/6		0.11	0.14	0.10
CE	2	1	1/2	→	0.22	0.29	0.30
EC	6	2	1		0.67	0.57	0.60
Column total	9.00	3.50	1.67		$\Sigma = 1$	$\Sigma = 1$	$\Sigma = 1$

Table 5.33   Derivation of initial estimates of decision criteria weights using approximation method.

Criterion	Calculation	Result
RC	$(0.11 + 0.14 + 0.10) \div 3$	0.118
CE	$(0.22 + 0.29 + 0.30) \div 3$	0.269
EC	$(0.67 + 0.57 + 0.60) \div 3$	0.613
	$\Sigma =$	1.0000

The approximation method yields the initial set of weights in Table 5.33.

Two iterations of the exact method yield a stable score for the weightings of each of the main criteria (Tables 5.34 and 5.35).

The final normalised weights are detailed within Table 5.36.

Table 5.34   Derivation of decision criteria weights using exact method – step 1.

1	1/2	1/6		0.118		0.355		0.117
2	1	1/2	×	0.269	=	0.812	Normalised →	0.268
6	2	1		0.613		1.859		0.615

Table 5.35   Derivation of decision criteria weights using exact method – step 2.

1	1/2	1/6		0.117		0.354		0.117
2	1	1/2	×	0.268	=	0.810	Normalised →	0.269
6	2	1		0.615		1.855		0.614

Table 5.36 Final normalised criterion weightings.

RC	=	0.12
CE	=	0.27
EC	=	0.61

Determining the relative performance of the six options on each of the decision criteria.

The relative performance of the six options on the basis of the three decision criteria are detailed in Table 5.37.

In all cases, low scores are best. On the basis of these performances, using the AHP scale, the relative performances of the six options on each of the three decision criteria are given within Tables 5.38–5.40.

Taking the RC criterion first, the approximation method is used to obtain an initial set of average scores for the six options, as shown in Table 5.41.

Two iterations of the exact method, as detailed in Tables 5.42 and 5.43, yield a stable score for the weightings of each of the main criteria.

The final normalised scores for the six options on the RC criterion are detailed in Table 5.44.

Table 5.37 Numerical values of various criteria for all options (lowest is best).

Options	Criterion scores		
	RC	CE	EC
A_1	3.3	8.3	206
A_2	4.1	7.5	198
A_3	3.7	7.3	200
A_4	3.7	7.3	191
A_5	1.8	8.9	216
A_6	1.7	8.3	199

Table 5.38 Relative performance of all options on the RC criterion.

RC	A_1	A_2	A_3	A_4	A_5	A_6
A_1	1	2	2	2	1/3	1/4
A_2	1/2	1	1/2	1/2	1/5	1/6
A_3	1/2	2	1	1	1/4	1/5
A_4	1/2	2	1	1	1/4	1/5
A_5	3	5	4	4	1	1
A_6	4	6	5	5	1	1

Table 5.39   Relative performance of all options on the CE criterion.

CE	A_1	A_2	A_3	A_4	A_5	A_6
A_1	1	1/2	1/3	1/3	3	1
A_2	2	1	1	1/2	5	3
A_3	3	1	1	1	6	5
A_4	3	2	1	1	6	5
A_5	1/3	1/5	1/6	1/6	1	1/3
A_6	1	1/3	1/5	1/5	3	1

Table 5.40   Relative performance of all options on the EC criterion.

EC	A_1	A_2	A_3	A_4	A_5	A_6
A_1	1	1/3	1/3	1/5	4	1/3
A_2	3	1	1	1/3	5	1
A_3	3	1	1	1/3	5	1
A_4	5	3	3	1	7	3
A_5	1/4	1/5	1/5	1/7	1	1/5
A_6	3	1	1	1/3	5	1

Table 5.41   Average set of weights derived for RC criterion using approximation method.

RC	A_1	A_2	A_3	A_4	A_5	A_6	A_1	A_2	A_3	A_4	A_5	A_6	Average
A_1	1	2	2	2	1/3	1/4	0.11	0.11	0.15	0.15	0.11	0.09	**0.12**
A_2	1/2	1	1/2	1/2	1/5	1/6	0.05	0.06	0.04	0.04	0.07	0.06	**0.05**
A_3	1/2	2	1	1	1/4	1/5	0.05	0.11	0.07	0.07	0.08	0.07	**0.08**
A_4	1/2	2	1	1	1/4	1/5	$\rightarrow$ 0.05	0.11	0.07	0.07	0.08	0.07	**0.08**
A_5	3	5	4	4	1	1	0.32	0.28	0.30	0.30	0.33	0.36	**0.31**
A_6	4	6	5	5	1	1	0.42	0.33	0.37	0.37	0.33	0.36	**0.36**
Total	9.5	18	13.5	13.5	3.03	2.82	1.00	1.00	1.00	1.00	1.00	1.00	

Table 5.42   Derivation of option scores on RC criterion using exact method – step 1.

1	2	2	2	1/3	1/4		0.1186	0.7260	0.1188
1/2	1	1/2	1/2	1/5	1/6		0.0512	0.3110	0.0509
1/2	2	1	1	1/4	1/5	$\times$	0.0776 =	0.4675 Normalised $\rightarrow$	0.0765
1/2	2	1	1	1/4	1/5		0.0776	0.4675	0.0765
3	5	4	4	1	1		0.3118	1.9073	0.3121
4	6	5	5	1	1		0.3633	2.2322	0.3653

Table 5.43   Derivation of option scores on RC criterion using exact method – step 2.

1	2	2	2	1/3	1/4		0.1188	0.7219		0.1186
1/2	1	1/2	1/2	1/5	1/6		0.0509	0.3101		0.0510
1/2	2	1	1	1/4	1/5	×	0.0765 =	0.4652	Normalised →	0.0765
1/2	2	1	1	1/4	1/5		0.0765	0.4652		0.0765
3	5	4	4	1	1		0.3121	1.9000		0.3122
4	6	5	5	1	1		0.3653	2.2227		0.3653

Table 5.44   Final normalised option scores on the RC criterion.

A_1	=	0.12
A_2	=	0.05
A_3	=	0.08
A_4	=	0.08
A_5	=	0.31
A_6	=	0.36

Table 5.45   Final normalised option scores on the CE criterion.

A_1	=	0.10
A_2	=	0.20
A_3	=	0.28
A_4	=	0.31
A_5	=	0.04
A_6	=	0.08

The exact method yields a stable set of performance scores for each option on the CE criterion as detailed in Table 5.45.

The exact method yields a stable set of performance scores for each option on the CE criterion, as detailed in Table 5.46.

The scores for the six options on each of the decision criteria are multiplied by the derived weightings in order to obtain the final overall performance scores as detailed in Table 5.47.

A_4(34%) is ranked first, with A_3(18%) a distant second, followed by A_2 and A_6 (both 16%).

Table 5.46   Final
normalised option
scores on the EC
criterion.

A_1	=	0.07
A_2	=	0.16
A_3	=	0.16
A_4	=	0.40
A_5	=	0.03
A_6	=	0.16

Table 5.47   Estimation of overall performance of the six options.

	RC	CE	EC		Weight		Score
A_1	0.12	0.10	0.07		0.12		0.08
A_2	0.05	0.20	0.16	×	0.27	=	0.16
A_3	0.08	0.28	0.16		0.61		0.18
A_4	0.08	0.31	0.40				0.34
A_5	0.31	0.04	0.03				0.07
A_6	0.36	0.08	0.16				0.16

5.5   Concordance analysis

5.5.1   Introduction

Concordance analysis is a 'partially compensating' decision modelling technique. It is described as such because there is no question of the 'trading-off' of
one criterion directly against another for each project option so that each can
be given a cumulative score indicating its 'attractiveness' (Rogers and Bruen,
1995). It is a pairwise method where, on each criterion, the level of dominance of one given option over another determines its concordance score. For
every pair of options, the scores on the different criteria are combined, using
the importance weightings chosen for the different criteria, to give a concordance index for that pair. As this can be seen as an indirect form of 'trading-off'
scores from one criterion against another, the process is called 'partial compensation'. The end result is a ranking or grouping of the options concerned
rather than the determination of a score for each. Calculation of the criterion
weightings is again a very important step within this type of model.

The simple additive weighted and AHP models result in the formation
of a function allowing the ranking from best to worst of all project options.
The information obtained is thus quite comprehensive, with a cardinal score
derived for each element in question. However, this richness of information

is a direct consequence of the high quality of data required to be input into it, data that may not be readily to hand if the engineering evaluation of the various options is preliminary in nature. Furthermore, the decision maker may not actually require the high level of information provided by the additive model. What may be required is a decision method that generates a simple ranking of the options involved rather than an actual score for each. In such a situation, concordance analysis may be the most appropriate decision model. Furthermore, it is a suitable methodology in situations where it may not be necessary to know the relative positions in the final hierarchy of all options. For example, if it is known that option a is better than options b and c, it may be irrelevant to the decision maker what the relative position of b and c is. It should be possible for the two to remain incomparable without endangering the decision process. In some engineering situations, it might actually be quite useful to highlight the incomparability of a number of options because of the absence of sufficient information to allow any meaningful comparison. In such circumstances, the result will reflect a more realistic solution given the quality of data available. If the information is not of a sufficiently high quality to produce a ranking directly connecting all options, then what is termed a 'partial ranking' will be derived where some options are not directly compared. In practice, concordance techniques are generally used to produce a shortlist of preferred options from a relatively large number of project options rather than one single 'best' project alternative.

With concordance analysis, the comparison of project options takes place on a pairwise basis with respect to each criterion and establishes the degree of dominance that one option has over another. The main measure of this dominance of one option over another is the concordance score. For a given pair of options (a,b), comparison of their relative performance takes place on each individual criterion. As a result, a picture of the level of dominance of a over b is built up.

Thus

If option a is at least as good as option b on criterion j, then

$$C_j(a, b) = 1.0$$

If a is **not** at least as good as b on criterion j, then

$$C_j(a, b) = 0$$

This exercise is done on (a, b) for each criterion $j, j = 1, n$.

Say there are five criteria, $j = 1,5$, and the score for each option on these are given in Table 5.48.

On criterion 1, option a is at least as good as option b; therefore,

$$C_1(a, b) = 1.0$$

Table 5.48 Option scores on each criterion (concordance calculation).

	Criterion 1	Criterion 2	Criterion 3	Criterion 4	Criterion 5
Option a	4	3	4	2	4
Option b	4	4	2	2	3

On criterion 2, option a is not at least as good as option b; therefore,

$$C_2(a, b) = 0$$

On criterion 3, option a is at least as good as option b; therefore,

$$C_3(a, b) = 1.0$$

On criterion 4, option a is at least as good as option b; therefore,

$$C_4(a, b) = 1.0$$

On criterion 5, option a is at least as good as option b; therefore,

$$C_5(a, b) = 1.0$$

It should be noted that the minimum data required to compile a set of concordance scores on a given criterion is a ranking of the options on that criterion. This can form the basis for all decisions regarding dominance.

Now that we have compiled the concordance scores for all the criteria in question, how do we combine these scores to give an overall indication of the dominance of b by a? We achieve this by use of the relative importance weightings for the criteria.

They are obtained using one of the weighting procedures outlined in Section 5.3.3. Use of such methods might result in the scores given in Table 5.49.

The overall concordance index for (a, b) is obtained by multiplying each criterion concordance score by its normalised weight.

$$C(a, b) = 0.125(1) + 0.25(0) + 0.25(1) + 0.125(1) + 0.125(1)$$

$$= 0.75$$

Table 5.49 Criterion weights.

Criterion	Weighting score	Normalised weight
1	1	0.125
2	2	0.250
3	2	0.250
4	1	0.125
5	2	0.125

Table 5.50   Sample concordance matrix.

	1	2	3	4	5	6
1		0.75	0.50	0.63	0.13	0.25
2	0.63		0.38	0.63	0.25	0.25
3	0.50	0.63		0.75	0.13	0.25
4	0.50	0.50	0.38		0.25	0.25
5	1.00	0.75	1.00	0.75		0.50
6	1.00	1.00	0.75	0.75	0.63	

The nearer this value is to 1, the more certain we are that a outranks or dominates b.

This index is calculated for each pair of options to form the concordance matrix.

The concordance index can be expressed in mathematical form in Table 5.50.

Assuming each criterion is assigned a weight w_j ($j = 1, n$; $n =$ number of criteria), increasing with the importance of the criterion, the concordance index for each ordered pair (a, b) can be rewritten as follows:

$$C(a, b) = \frac{1}{W} \sum_{j: g_j(a) \geq g_j(b)} w_j \tag{5.4}$$

where

$$W = \sum_{j=1}^{n} w_j \tag{5.5}$$

and $g_j(a)$ is the score for criterion j under option a and and $g_j(b)$ is the score for criterion j under option b.

$C(a, b)$ has a value between 0 and 1 and measures the strength of the statement 'Project option A outranks Project option B'.

5.5.2   PROMETHEE I

Now that the pairwise dominance relationship between options has been established by means of the indices within the concordance matrix, one must now go about establishing a relative ranking of the proposals based on these dominance scores. One of the most straightforward ways of compiling the ranking hierarchy is the system put forward by Brans and Vincke (1985) as part of their PROMETHEE I decision method.

The process put forward by Brans and Vincke is carried out for each pair of n options. Assume first that all concordance indices have been calculated and the concordance matrix has been compiled. Let us take a sample concordance matrix with six options compared pairwise as follows.

For a given option, the sum of the scores along its row indicates the extent to which it is better than all the other options, while the sum of the scores along its column is an indication of the degree to which the other options are better than it. Therefore, in the ideal case, an option being compared against four others would have a row score of 4 and a column score of 0. The greater the options row score and the smaller its column score, the better its ranking.

PROMETHEE I uses this property by taking each option's row score and ranking each option, placing the one with the highest score first, the one with the next highest score second and so on until all are ranked. The result is a *complete pre-order*, as all options are directly connected within the ranking structure, with ties between options allowed (it would be a complete *order* if ties were not permitted). The method then takes the column scores and places the option with the highest score last, the one with the next highest score second last and so on. The result is again a complete pre-order. A final ranking is obtained by finding the intersection between these two separate rankings. This final result can yield a *partial pre-order*, as incomparabilities between options may occur due to possible conflicts between the two rankings.

Example 5.12 Ranking Six Alternative Renewable Energy Sources Using PROMETHEE I.

Taking the baseline data from Example 5.4, six options are proposed for developing renewable energy sources within a region, wind (W), biomass (B), photovoltaic (PV), a combination of wind–biomass (WB), a combination of wind–PV (WPV) and a combination of wind–biomass–PV (WBP).

Table 5.51 details the relative rankings of the six options on the five decision criteria, based on their scores within Table 5.6.

Table 5.51 Ranking of the six options on the five decision criteria.

Employment	Land used	Visual	Emissions	Efficiency
W,WB,WBP	B,WB,WPV, WBP	B	W	W
↓	↓	↓	↓	↓
B,WPV	W	PV,WB	PV,WB,WPV	PV,WPV
↓	↓	↓	↓	↓
PV	PV	W,WPV,WBP	B,WBP	WB,WBP
				↓
				B

The same weightings as derived within Example 5.7 are utilised within this example:

Land used: 0.30
Employment: 0.25
Emissions: 0.20

Visual: 0.15

Efficiency: 0.10.

On the basis of both the derived rankings and the importance weightings assigned to each criterion, a set of concordance indices for each pair of options on each of the eight decision criteria were compiled. These are obtained by multiplying the set of concordance scores on each criterion by its normalised weighting. These are shown in Tables 5.52–5.56. For each pair of options, the concordance index on each criterion is summed to give an overall concordance index. These values are shown in Table 5.57. The row sum and column sum scores for each option are given in Table 5.58, and the final partial pre-order obtained from the intersection of the two rankings from Table 5.58 are detailed in Figure 5.2.

Table 5.52   Concordance scores of the six options on the employment criterion.

	W	B	PV	WB	WPV	WBP
W		0.25	0.25	0.25	0.25	0.25
B	0		0.25	0	0.25	0
PV	0	0		0	0	0
WB	0.25	0.25	0.25		0.25	0.25
WPV	0	0.25	0.25	0		0
WBP	0.25	0.25	0.25	0.25	0.25	

Table 5.53   Concordance scores of the six options on the land used criterion.

	W	B	PV	WB	WPV	WBP
W		0	0.3	0	0	0
B	0.3		0.3	0.3	0.3	0.3
PV	0	0		0	0	0
WB	0.3	0.3	0.3		0.3	0.3
WPV	0.3	0.3	0.3	0.3		0.3
WBP	0.3	0.3	0.3	0.3	0.3	

Table 5.54   Concordance scores of the six options on the visual criterion.

	W	B	PV	WB	WPV	WBP
W		0	0	0	0.15	0.15
B	0.15		0.15	0.15	0.15	0.15
PV	0.15	0		0.15	0.15	0.15
WB	0.15	0	0.15		0.15	0.15
WPV	0.15	0	0	0		0.15
WBP	0.15	0	0	0.15	0.15	

Table 5.55   Concordance scores of the six options on the emissions criterion.

	W	B	PV	WB	WPV	WBP
W		0.2	0.2	0.2	0.2	0.2
B	0		0	0	0	0.2
PV	0	0.2		0.2	0.2	0.2
WB	0	0.2	0.2		0.2	0.2
WPV	0	0.2	0.2	0.2		0.2
WBP	0	0.2	0	0	0	

Table 5.56   Concordance scores of the six options on the efficiency criterion.

	W	B	PV	WB	WPV	WBP
W		0.1	0.1	0.1	0.1	0.1
B	0		0	0	0	0
PV	0	0.1		0.1	0.1	0.1
WB	0	0.1	0		0	0.1
WPV	0	0.1	0.1	0.1		0.1
WBP	0	0.1	0	0.1	0	

Table 5.57   Overall concordance matrix for the six options.

	W	B	PV	WB	WPV	WBP	Row sum
W	0	0.55	0.85	0.55	0.7	0.7	**3.35**
B	0.45	0	0.7	0.45	0.7	0.65	**2.95**
PV	0.15	0.3	0	0.45	0.45	0.45	**1.8**
WB	0.7	0.85	0.9	0	0.9	1	**4.35**
WPV	0.45	0.85	0.85	0.6	0	0.75	**3.5**
WBP	0.7	0.85	0.55	0.8	0.7	0	**3.6**
Column sum	**2.45**	**3.4**	**3.85**	**2.85**	**3.45**	**3.55**	

Table 5.58   Rankings based on row sum and column sum scores.

	Row sum	Position	Column sum	Position
W	3.35	Fourth	2.45	First
B	2.95	Fifth	3.40	Third
PV	1.80	sixth	3.85	Sixth
WB	4.35	First	2.85	Second
WPV	3.5	Third	3.45	Fourth
WBP	3.60	Third	3.55	Fifth

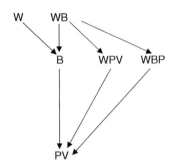

Figure 5.2   Overall ranking of the six options using PROMETHEE I.

One can see the result is different from that obtained using the SAW method. The wind–biomass option, which was marginally second using SAW, is first in this instance, with the wind option also performing well. The biomass option, which was ranked marginally first using SAW, is ranked third in this instance, below wind and wind–biomass.

5.5.3 ELECTRE TRI

ELECTRE TRI (Yu, 1992a, 1992b) is a multiple criteria sorting method, that is, a method that assigns alternatives to pre-defined categories. The assignment of an alternative a results from the comparison of a with the profiles defining the limits of the categories. Let F denote the set of the indices of the criteria $g_1, g_2, \ldots, g_m$. ($F = \{1, 2, \ldots, m\}$) and B the set of indices of the profiles defining $p + 1$ categories ($B = \{1, 2, \ldots, p\}$), b_h (a reference option) being the upper limit of category C_h and the lower limit of category C_{h+1}, $h = 1, 2, \ldots, p$.

ELECTRE TRI assigns options to categories based on the construction of an outranking relation S that characterises how options compare to the limits of categories. The model builds an outranking relation by validating/invalidating the assertion aSb_h, on the basis that a is at least as good as the reference option b_h. If the option a outranks b_h but does not outrank b_{h+1}, the option is thus placed within the category spanning the limits b_h and b_{h+1}.

The following rules apply to the application of ELECTRE TRI to a decision problem:

- Each option can be assigned to one category only.
- The assignment of an option to its allotted category is not dependent on the assignment of any of the other options.
- When two options have the same outranking relationship with a given reference option, they must be assigned to the same category.
- If option b outranks option a, then option b must be assigned to a category at least as good as the one to which a is assigned.

Section 5.6 contains a case study which uses the ELECTRE TRI model in its basic form.

5.6 Site selection for wind farms – a case study from Cavan (Ireland)

5.6.1 Introduction

Cavan is an inland county in the province of Ulster, covering an area of 189,060 ha. Its population is approximately 60,000, living in mostly rural locations, with less than 20% of its population living in the three towns of 1500 and over. Its topography is characterised by drumlin countryside dotted with many lakes and small hills. The northwest of the county is particularly sparsely populated and mountainous. In general, the county has several areas of high scenic landscape, and overall has a pleasing pastoral environment. The county is, on paper, suitable for wind farm development.

The case study involves taking 18 potential wind farm sites in County Cavan and using MCDA to assess their overall suitability for development. The method of assessment relies both on national guidelines, those published by individual local authorities within Ireland, and relevant research carried out internationally (Rogers et al. (2007)).

5.6.2 National and international guidance

The Irish Department of Environment, Heritage and Local Government have published a draft copy of *Wind Farm Planning Guidelines* to assist the proper planning of wind power projects in appropriate locations around Ireland. The guidelines, first published in 1996, have been recently revised to account for changes in renewable energy policy, modern wind turbine technology evolution and amended procedures to encourage wind energy at a localised planning level.

The Department of Environment, Heritage and Local Government, within its Draft Wind Farm Planning Guidelines, states that a number of different studies can be undertaken to classify a county's landscape according to sensitivity to wind farm development. These may involve to varying degrees the public, the local planning authority and/or the landscape consultants. An outline of a stepwise procedure is provided as follows:

Desk reviews. Involves carrying out a literature review of the most pertinent documents relating to landscape sensitivity, including the current county development plan.

Consultations with planning staff. Involve carrying out an exercise to map landscape quality with the aid of landscape consultants, if considered appropriate, providing planning staff with large-scale maps of the county to map the following, using their own experience and impressions:

(a)   Those scenic routes which are currently designated in the development plan but which might have deteriorated over time such that they no longer warrant designation;

(b)   Roads currently not designated in the local authority's development plan as scenic routes but which are perceived to be of such high quality that they might warrant designation;

(c)   Areas of landscape that are of such high quality that they could not accommodate wind farms; and

(d)   Areas of landscape that could accommodate wind farm development.

Initial field work. Planning staff and/or landscape consultants spend time in the field examining the landscapes and identifying those locations that are deemed to be of very high quality. Areas of high quality will tend to agree with those highlighted in the mapping exercise carried out above.

Public consultation. At least one open focus group meeting (perhaps two or more in larger counties) should be held in order to consult with local communities and other interested parties regarding the development of a wind energy strategy for the county. These events should be well advertised and all statutory consultees should be formally invited. Following introductory presentations by the planning staff and/or consultants, the focus group attendees can participate in a series of hard-copy mapping exercises with the aim of identifying

(a)   Landscapes of exceptionally high quality;

(b)   Ordinary (non-remarkable) landscapes;

(c)   Locations where wind farm development would be unacceptable (to the consultees); and

(d)   Locations where wind farms could be acceptable.

Such mapping exercises are a very effective medium for classifying landscape sensitivity according to different values held by the attendees. Despite the wide ranging views held by different individuals and groups (ranging from very pro-development to highly conservative), common ground will often be found when the maps are collectively reviewed. The focus group classification of landscape sensitivity to wind farm development will often closely agree with the findings from the consultation with the planning authority, as well as with the conclusions drawn from the initial field work by the planning staff and/or landscape consultants.

Preparation of draft sensitivity map. All the marked-up maps prepared by the focus groups are examined in detail and a broad classification of sensitivity can be developed using geographic information systems (GIS). As an example, areas which are identified as being exclusion zones by both the planning authority and the focus groups and which were assessed as being of high quality might be classified as being high in sensitivity. Conversely, areas where little concern was expressed regarding wind farms, and where the consultants felt that quality was unremarkable, could be classified as being moderate, or in some cases low, sensitivity. The sensitivity classification could include areas that are

(a) acceptable in principle;
(b) open to consideration; and
(c) not acceptable.

A number of Irish local authorities have expanded on this classification system in recent times in order to be in a position to rank the desirability of a potential wind farm site.

Clare County Council published their Wind Energy Strategy as part of their 2011–2017 county development plan. On the basis that particular attention needed to be paid to identifying strategic sites of regional and national interest, four ranking levels, as shown in Table 5.59, were derived with criteria detailed to allow sites to be allocated to their appropriate place in the hierarchy.

The following are the detailed criteria used as a basis for allocating a site to one of the four ranking levels:

Strategic areas
- ○ Viable wind speeds;
- ○ Proximity to power grid;
- ○ Ground slope less than 15°;
- ○ Excludes all Special Areas of Conservations (SACs), Special Protection Areas (SPAs) and Natural Heritage Areas (NHAs); and
- ○ At least 400 m from residential and commercial properties.

Areas that are 'acceptable in principle'
- ○ Viable wind speeds;
- ○ Proximity to power grid;
- ○ Ground slope less than 15°;
- ○ Excludes all SAC's and SPA's and most NHA's; and
- ○ Low population densities.

Areas that are 'open for consideration'
- ○ Viable wind speeds;
- ○ Proximity to power grid; and
- ○ Dense pockets of population within area.

Areas that are 'not normally permissible'
- ○ Contains large number of heritage designations or important recreational/tourism areas;

Table 5.59   Four ranking levels.

Strategic areas
↑
Areas acceptable in principle
↑
Areas open for consideration
↑
Areas not normally permissible

    o   Contains large number of archaeological sites; and

    o   Strategic environmental assessment has recommended against inclusion of these areas.

Galway County Council, in their 2011 Wind Energy Strategy, derived a very similar classification, as shown in Table 5.60.

The following are the detailed criteria used as a basis for allocating a site to one of the five ranking levels:

Strategic areas Larger areas in most suitable locations for wind farm development without significant environmental constraints, based on strategic-level analysis. Wind farm developments to be encouraged in these areas subject to detailed environmental and visual assessment and appropriate layout and design.

Areas that are 'acceptable in principle' Smaller areas in suitable locations for wind farm development and without significant environmental constraints, based on strategic-level analysis. Wind farm developments will be facilitated in these areas subject to detailed environmental and visual assessment and appropriate layout and design

Areas that are 'open for consideration' Areas with some locations that may have potential for wind farm development due to viable wind speeds or clustering with strategic areas but with significant environmental constraints, based on strategic-level assessment. Wind farm developments in these areas will be evaluated on a case by case basis subject to viable wind speeds, environmental resources and constraints and amenity, safety and cumulative impacts

Areas that are 'not normally permissible' Areas generally not suitable for wind farm development due to their overall sensitivity and constraints arising from landscape, ecological, recreational, settlement, infrastructural and/or cultural and built heritage resources, based on strategic level assessment. Wind farm developments in these areas will be discouraged,

Table 5.60 Classifications derived by Galway County Council.

| Strategic areas |
| ↑ |
| Areas acceptable in principle |
| ↑ |
| Areas open for consideration |
| ↑ |
| Areas not normally permissible |
| ↑ |
| Areas of low wind speed |

unless project-level habitats directive assessment (HDA) and environmental impact assessment (EIA) can demonstrate to the satisfaction of the planning authority that environmental and other impacts can be successfully avoided, minimised and/or mitigated.

Low wind speed areas Areas with wind speeds less than 8 m/s would generally not provide viable locations for commercial wind farm developments.

Internationally, research carried out at the University of Dortmund and the University of Bochum (Therivel and Partidario, 1996) led to the development of an strategic environmental assessment framework study for wind power developments in the Soest district, Germany. The process proactively steers investment in wind farms in the district by establishing criteria for an ecologically oriented, land-use-based approach to wind farm sitting.

The process of deciding where wind farms could be located was based both on incident wind speed and the categorisation of the site into one of the three following ranking levels:

Excluded areas Areas that would not be available for wind farm developments for reasons of their designations, use or value

Restricted areas Where wind farms are not unreservedly supported and must be environmentally assessed through a short report

Favoured areas Where no landscape or ecological concerns exist, and where economic conditions, such as sufficient wind speeds, are met.

The assessment of the areas concerned involve identifying conflicts between wind power generation and the land use designations in the planning authority's development plans, conservation areas and ecologically valuable areas.

Table 5.61 details some of the criteria used to screen the sensitivity of different locations within the Soest district regarding wind power uses.

Moreover, in the Castilla y Leon region of Spain, research carried out in 2001 indicated that wind farm site selection was based on a decision-making process using both quantitative and qualitative methods.

Table 5.61   Criteria for selection of wind farm sites.

Suitability criteria	Exclusionary criteria	Restriction criteria
Average annual wind speed of 5 m/s at 10 m height and possibly areas with 4.5 m/s annual wind speed	Within nature conservation areas or national parks	Areas protected by nature conservation law
	100 m from electricity transmission corridors	
	50 m from transmission lines of $\geq$30 kV	
	$\geq$200 m from forests	

Source: Data taken from multiple sources by Therivel and Partidario, 1996.

Using GIS, a composite map was generated for each of the following five elements:

- Natural spaces;
- Visual impact;
- Vegetation and associated fauna biotopes;
- Socio-economic impacts and historical–cultural and archaeological heritage; and
- Areas of geomorphological risk.

Each map divided the region into four levels of environmental impact; 1 being negligible and 4 being extremely high.

A final map was assembled showing an overall level of environmental impact, again graded on the aforementioned four-point scale, with the overall value for environmental sensitivity being defined by a weighting of each of the five elements, with natural spaces and visual impacts receiving the heaviest weightings. Technical feasibility for the different areas within the region was also graded from 1 to 4 (very high, high, sufficient and low).

Four alternative outcomes were defined depending on the degree of environmental protection and the level of technical feasibility. A matrix was formed in which, for each area, the degree of protection was selected:

Alternative Action 1. Unfeasible development – wind farms not allowed.

Alternative Action 2. Limited development – areas of land with this classification would be subject to full EIA, with stringent mitigation and monitoring measures required (e.g. requiring a minimum distance between adjacent wind farms in order to minimise synergistic impacts).

Alternative Action 3. Controlled development where full EIA required, with some mitigation and monitoring measures required.

Alternative Action 4. Free development – limited EIA required.

5.6.3   Decision framework chosen

The method used to classify the 18 potential wind farm locations in County Cavan on the basis of their suitability for wind farms is derived from a combination of all the decision systems described above. Like Castilla, the Draft Irish Wind Farm Guidelines and the published strategy for both Cavan and Galway, four basic rankings for potential sites are used, with the criteria used to determine the suitability rating of each site taken from the Soest (suitability, exclusionary and restriction criteria).

In order to assess and rate various potential within the county area, the following rating system is put forward:

Category 1 – Extremely Sensitive Location A wind farm should not be considered on a site with this designation. A number of major factors exist that restrict development on the site. Any area deemed an SAC or an NHA would be seen as a key factor in its classification.

Category 2 – Sensitive Location This designation implies a site which is highly restrictive to wind farm development. Special heritage sites, high landscape areas and scenic viewing points fall under this heading. Advanced levels of wildlife will also imply this classification. As wind farm development involves a considerable amount of disruption to the area, in the form of site clearance, construction of access roads and pipe laying, the local wildlife is seen to suffer significantly as a result of these construction activities.

Although sensitive areas do not by definition preclude wind farm development, it is generally considered unfavourable at such a location. Permission should only be granted after an in-depth assessment of the environmental impact of the proposal.

Category 3 – Suitable Location Sites designated as suitable can be assumed to be favoured regions for wind farm development. They possess no heritage or conservation restrictions. They may, however, be close to private housing and suburban areas. Wind farm development, while favoured, should be regulated and assessed through EIA.

Category 4 – Highly Suitable Location This designation implies a site that is extremely suitable as a wind farm. In addition to having low populations in their vicinity, further characteristics which would further reinforce their suitability would include proximity to high voltage lines, the availability of a large land area and the absence of masts, particularly those for television broadcasting.

All of these factors outlined in Cavan County Council's Development Plan relating to the selection process will be taken into account within this assessment.

Using the onshore wind atlas (OWA) (Sustainable Energy Ireland, 2003) to identify all the sites for assessment, the basic criteria was that the site must have a minimum on-site wind speed of 7.5 m/s or greater. This value is seen by the energy industry within Ireland as an acceptable benchmark for economic viability.

Table 5.62 details the general location within County Cavan of the 18 sites, with Figure 5.3 containing a map indicating the location of each site within the county boundary.

Table 5.63 lists the performance on the various criteria that would determine which of the aforementioned four categories/designations a potential site will be allocated:

5.6.4   Decision model utilised to categorise each of the 18 sites

The decision model chosen for this case study is a simplified version of ELEC-TRE TRI.

Table 5.62   Definition of four possible designations for the site of a wind farm.

Site no.	Site name	Location
1	Corrakeelum	Northwest Cavan
2	Cuilagh-Aneirin Uplands	Northwest Cavan
3	Slieve Rushen	North Cavan
4	Bruse Hill	West Cavan
5	Ardkill More/Beg	Southeast of Cavan Town
6	Shantermon	East of Cavan Town
7	Aghalion	South Cavan
8	Carnalynch	South of Baileborough
9	Seefin	Northwest of Baileborough
10	Greaghettaigh	Northwest of Baileborough
11	Tievenass	Northwest of Baileborough
12	Bindoo and Greaghnacross	Northwest of Baileborough
13	Gartnaneane	Northeast of Baileborough
14	Taghart South	South of Baileborough
15	Corkish	East of Baileborough
16	Slieve Glah	East of Cavan Town
17	Cornasaus	East Cavan
18	Ratrussan	Northwest of Baileborough

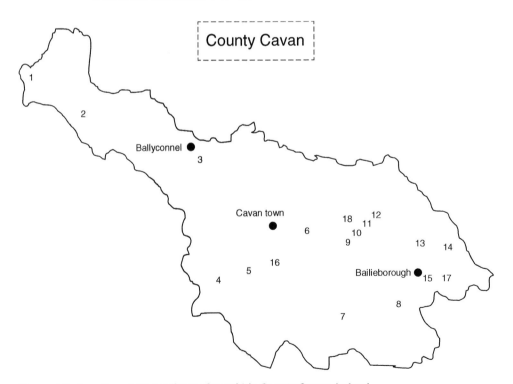

Figure 5.3   Location of 18 candidate sites within County Cavan, Ireland.

Table 5.63   Definition of four possible designations for the site of a wind farm.

	Exclusionary criteria	Restriction criteria	Suitability criteria
Extremely sensitive	Within an SAC NHA Very fragile/sensitive landscape (Cr1)		
Sensitive		Special heritage site/Special protection area or High landscape areas (Cr2) Scenic viewing points (EIA required) (Cr3) Major effect on local wildlife or natural habitat for major species (Cr4)	
Suitable		Close to private housing and suburban areas (Cr5) Limited by their land area (Cr6)	
Highly suitable			Close to high voltage lines (Cr7) Medium to large land area available (Cr8) Absence of TV masts in the vicinity (Cr9) No restrictions under County Council Development Plan (Cr10)

Abbreviations: SAC, Special Area of Conservation; HHA, National Heritage Area.

The ELECTRE-TRI method (Yu, 1992a, 1992b) belongs to the ELEC-TRE family of multi-criteria methods developed by Bernard Roy and his co-workers (Roy, 1968, 1978). This specific method is dedicated to the sorting problem: to assign each alternative to one of a set of predefined ordered categories according to a set of evaluation criteria. The categories (Cx) are defined by specifying their boundaries by means of reference profiles. ELECTRE methods rely on the construction and the exploitation of an outranking relation in the face of the problem to be tackled (selection, ranking or assignment). To say that 'alternative a outranks alternative b' means that 'a is at least as good as b'. The assignment of each action to one category is done by comparing its performances in each criterion to the performances of the reference profiles. The procedure assigns each action to the highest category such that its lower bound is outranked by a.

The model gives a methodological basis for the assignment of all 18 sites to one of the four categories detailed in Table 5.64.

Table 5.64   Assignment of option to one of four categories based on performance on certain criteria.

Criteria		Cat 1	Cat 2	Cat 3	Cat 4
Within SAC or NHA or very fragile/sensitive landscape or special heritage site/special protected area	Cr1	Y	N	N	N
Special heritage/high landscape areas	Cr2		Y	N	N
Scenic viewing points requiring EIA	Cr3		Y	N	N
Major effect on local wildlife or natural habitat for major species	Cr4		Y	N	N
Close to private housing and suburban areas	Cr5			Y	N
Land area limitations	Cr6			Y	N
Close to high voltage lines	Cr7			N	Y
Medium/large land area available	Cr8			N	Y
No TV masts in vicinity	Cr9			N	Y
No restrictions under county development plan	Cr10			N	Y

Table 5.65 indicates how the qualitative information for each site for the decision criteria in question results in the assignment of a candidate site to one of the four categories.

Only sites within categories 3 and 4 are considered suitable for development, with particular emphasis on those within category 4.

5.6.5   Selection of potentially suitable sites

Table 5.65 gives some detail of the qualitative information which resulted in the assignment of each of the site to one of the four categories.

One can see that proximity of a site to National Heritage Area or a Special Area of Conservation or being located within a very fragile/sensitive landscape has resulted in the assignment of the site to category 1 and its exclusion from further consideration. Sites with special heritage classifications, or with sensitive landscapes and/or scenic viewing points, were assigned to category 2 and also excluded from consideration. Those deemed suitable or highly suitable for development were assigned to category 3 or 4 depending on the level of planning or infrastructural limitations associated with them.

On the basis of the information in Table 5.65, a ranking of the 18 sites can be compiled. This ranking is detailed in Figure 5.4.

5.6.6   Concluding comment on case studies

Using County Cavan as an example, the above procedure demonstrates that, use of a simplified decision model to allow the planning authority to proactively categorise suitable locations for wind farms rather than passively react to planning applications for one-off particular sites within the planning

Table 5.65   Definition of four possible designations for the site of a wind farm.

Site name	Performance on evaluation criteria	On-site wind speed (m/s)	Ranking (four-point scale)
Corrakeeldrum	High landscape area within very fragile and sensitive landscape	9.5	Extremely sensitive
Cuilagh-Aneirin Uplands	National heritage area	8	Extremely sensitive
Slieve Rushen	Remote, with no restrictions in relation to heritage/conservation policies. No TV masts	8.25	Highly suitable
Bruse Hill	Natural habitat for a vast array of species	8	Sensitive
Ardkill More/Beg	Area suitable for small-scale housing development No environmental restrictions	8	Suitable
Shantermon	Remote with few houses special heritage status	8.25	Sensitive
Aghalion	Housing in the area No restrictions under CCDP	8.5	Suitable
Carnalynch	Surrounded by housing No restrictions under CCDP	7.75	Suitable
Seefin	Low population No restrictions under CCDP	7.75	Suitable
Greaghettaigh	No obstruction by any forestry Little housing in the vicinity Poor access to the national grid and poor road infrastructure	8	Suitable
Tievenass	Vastly unpopulated and generally remote No restrictions under CCCP Immediate access to a 110 kV transmission line. No TV masts.	8.5	Highly suitable
Bindoo and Greaghnacross	Exposed and very isolated area No restrictions under CCDP Nearby access to 110 kv line	8.25	Highly suitable
Gartnaneane	Development unlikely as region almost completely covered by existing wind farm	7.75	Sensitive
Taghart South	Few scattered houses nearby Nearby connection to the grid No restrictions under CCDP	8.25	Suitable
Corkish	Small population nearby Nearby access to 38 and 110 kV power lines No restrictions under CCDP	8	Suitable
Slieve Glah	Two 38 kV power lines in close proximity Small number of houses in the vicinity, some with large holdings No restrictions under CCDP	8.25	Suitable
Cornasaus	Nearby access to 38 and 220 kV power lines High landscape area under CCDP (fragile and sensitive landscape)	8	Sensitive
Ratrussan	Excellent access to national grid Very few houses in the vicinity No TV masts in vicinity No known animal habitats	7.75	Highly suitable

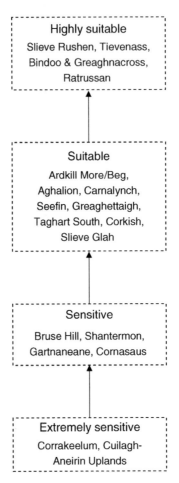

Figure 5.4   Ranking of 18 sites assessed on the basis of their classification within one of four categories.

authority's jurisdiction, represents an environmentally sustainable method of planning for this renewable energy source.

Sites selected by the existing method may have been chosen by the individual developer on different merits such as the low probability that the application will be objected to or the cost savings arising from the availability of a nearby grid connection.

5.7   Concluding comments on MCDA models

The simple non-compensatory models described in the chapter provide many of the basic elements required for a sound decision process. The data requirements in most cases are relatively modest; some only requiring detailed data on the most important criteria. Some, such as the lexicographic semi-order allow uncertainty regarding criterion valuations to be brought into the model by the use of thresholds. In the context of the appraisal of complex renewable energy projects, where all decision criteria need to be examined in some detail, use of the attitude-oriented or lexicographic methods, which concentrate on each option's performance on one sole criterion, may be inappropriate. Furthermore, the use of a dominance method to analyse a complex performance matrix containing options whose performances on the various criteria are conflicting would not yield worthwhile results, as it is unlikely that very many options would be dominated. The method would thus not partition the proposals effectively into a relatively small kernel of shortlisted proposals. Satisficing techniques tend to identify options that meet some minimum acceptable standard set by the decision maker rather than one or more proposals that perform best within the overall decision matrix. Nonetheless, these simple techniques provide an excellent basis for understanding many of the fundamental theoretical aspects of the more intricate models such as the SAW method, the AHP technique and concordance analysis.

The SAW model requires criteria of evaluation, originally measured on different scales, to be evaluated on a common basis and permits multiple weighting systems reflecting the views of different decision-making groups to be input into the process. Sensitivity analysis is seen to play a major role within this technique, allowing the robustness of the baseline results to be gauged.

The AHP provides a framework for calibrating a numerical scale and is useful in renewable energy problems where pure quantitative valuations and comparisons do not exist. The participation of a group of decision makers make it possible to assess trade-offs between diverse criteria or options at a given level of analysis. It can be used as an aid to the decision-making process and enables compromise to be attained from the various judgments reached.

The concordance model outlined in detail within this chapter, PROMETHEE I, is relatively complex. It assesses the strengths and weaknesses of the options on the basis of their pairwise concordance indices. Concordance models are best used where a relatively large number of options must be reduced to a shortlist. Despite its complexity, concordance techniques such as PROMETHEE have a number of advantages. They are rigorous yet adaptable techniques, allowing criteria measures on different scales to be evaluated within the same framework. Unlike the SAW and AHP models, the criterion valuations do not have to be transferred onto a common scale before the relative performances of the different options can be evaluated. The method does not force direct comparison between any two options if insufficient information exists to do so or if the information that does exist gives conflicting indications. Given their ability to handle data of mixed quality presented in various formats, they have proved to be particularly useful for the preliminary assessment of a wide range of project proposals. ELECTRE TRI, which places options in categories rather than ranking them, offers a useful variation on the PROMETHEE model and is outlined briefly.

References

Brans, J.P. & Vincke, P. (1985) A preference ranking organisation method. *Management Science*, **31**(6), 647–656.

Hinkle, D. (1965) The Change of Personal Constructs from the Viewpoint of a Theory of Construct Implications. Ph.D. Dissertation, Ohio State University.

Keeney, R.L. & Raiffa, H. (1976) *Decisions with Multiple Objectives*. John Wiley & Sons, Inc, New York.

Miller, G.A. (1956) The Magical Number Seven, Plus or Minus Two: Some Limits On Our Capacity For Processing Information. *The Psychological Review*, Vol. **63**, pp. 81–97. Sustainable Energy Ireland (1993). Onshore Wind Atlas. Sustainable Energy Ireland, Sligo, Ireland.

Rogers, M.G. & Bruen, M.P. (1995) Non-monetary based decision-aid techniques in EIA – an overview. *Proceedings of the Institution of Civil Engineers, Municipal Engineer*, **109**(June), 98–103.

Rogers M.G., and Bruen, M.P. (1998) A new system for weighting criteria within Electre. *European Journal of Operational Research*, Vol. **107**, No. 3, pp. 552–563.

Rogers, M.G., Rogers, M.M. & Magee, I. (2007) Developing a site selection process for wind farms using strategic environmental assessment. *Paper 14938, Proceedings of the Institution of Civil Engineers, Engineering Sustainability*, **160**(June), 79–86.

Roy, B. (1968) Classement et choix en presence de points de vue multiples (la methode ELECTRE). *Revue Informatique et Recherche Operationnelle*. 2e Annee, No. 8, pp. 57–75.

Roy, B. (1978) ELECTRE III: Un algorithme de classements fonde sur une representation floue des preferences en presence de criteres multiples. *Cahiers de CERO*, Vol. **20**, No. 1, pp. 3–24.

Saaty, T.L. (1977) A scaling for priorities in hierarchical structures. *Journal of Mathematical Psychology*, **15**, 234–281.

Saaty, T.L. (1980) *The Analytic Hierarchy Process*. McGraw Hill, New York.

Therivel, R. & Partidario, M.R. (1996) *The Practice of Strategic Environmental Assessment*. Earthscan Publications Ltd., London, UK.

Yu, W (1992a) Electre Tri: aspects methodologiques at manual d'utilisation. *Document de LAMSADE*, No. 74, Universite Paris-Dauphine, p. 80.

Yu, W., 1992b. Aide Multicrit_ere a la Decision Dans le Cadre de la Problematique du Tri: Concepts, Methodes et Applications. Doctoral Dissertation, Universite de Paris-Dauphine.

6 Policy Aspects

In this chapter, we explore the use of concepts introduced previously to provide information for evidence-based policy making. We begin by explaining the main motivations for current international energy policies with a particular emphasis on environmental sustainability. An overview of common energy policy objectives, targets and instruments are then given. Methods for estimating marginal abatement costs (MACs) and designing renewable subsidies are then discussed. The quantification of the societal benefits and equity impacts of projects and policies is then considered. Finally, two case studies are presented.

6.1 Energy policy context

According to the Intergovernmental Panel on Climate Change's Fifth Assessment Report, 'warming of the climate system is unequivocal' (IPCC, 2013), as evidenced by atmospheric, oceanic and polar temperatures not experienced for millennia. It is 'extremely likely' that this is the result of human activity, in particular activities which result in the emissions of greenhouse gases (GHGs): concentrations of carbon dioxide (CO_2), methane (CH_4) and nitrous oxide (N_2O) are now at levels not seen for almost a million years. The greatest contributor to global warming is CO_2 emitted in energy conversion processes (see Figure 6.1). Therefore, in order to reduce the rate of emissions and global warming, the energy policies of many countries give significant importance to the environmental sustainability of energy supply and end use. The challenge for policy makers is how to balance this objective with providing cost-competitive energy to an economy.

Renewable Energy and Energy Efficiency: Assessment of Projects and Policies, First Edition.
Aidan Duffy, Martin Rogers and Lacour Ayompe.
© 2015 John Wiley & Sons, Ltd. Published 2015 by John Wiley & Sons, Ltd.
Companion Website: www.wiley.com/go/duffy/renewable

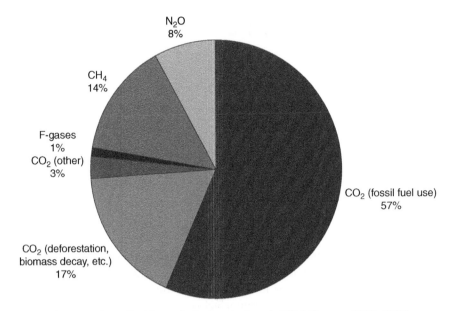

Figure 6.1   Global carbon dioxide equivalent emissions in 2004 (*Source:* IPCC, 2007).

Many mature low carbon energy supply and energy-efficient technologies are cost-effective to the private investor. For example, the cost of onshore wind has reached grid parity in many countries, depending on the available wind resource and the cost of electricity. In the United States, gas-fired power stations are displacing coal-fired plant in a cost-effective manner while at the same time reducing CO_2 emissions. However, the deployment potential of such cost-effective technologies may be insufficient to meet national or regional emissions' reduction targets and additional economic incentives may be required.

The economic theory of efficient markets suggests that the most cost-effective way to reduce emissions is to put an appropriate price on them. The market will identify least-cost technologies for emissions mitigation, thus minimising costs to society. One such example is carbon tax, which puts a credible, long-term price on emissions, thus increasing the cost of energy systems which emit relatively high quantities of CO_2. This incentivises the use of lower emissions technologies and processes, thus helping to meet national emissions targets.

However, supplementary policies are used in addition to carbon pricing in many countries for a number of reasons. Markets may not be perfect for a variety of reasons: there may be imperfect information or barriers to the adoption of cost-effective alternatives. Policies may be needed to counteract the lock-in of high emission technologies such as centralised fossil-fuelled power plant or to improve the political acceptability of desired changes in affected markets (Hood, 2011).

Therefore, policy makers also adopt further policies to encourage the uptake of low emission energy technologies. Informational policies are used in an attempt to overcome market failure, such as when cost-effective measures are underutilised. Such policies include energy-rating schemes for domestic appliances. Regulatory policies set minimum standards to overcome failures such as split incentives. For example, builders may wish to build houses without adequate insulation to save on construction costs, whereas owners would like a house that can be heated cost-effectively; policies on minimum building standards can solve this problem. Technology subsidies provide capital, taxation or operational incentives to private investors to encourage the deployment of specific technologies. Examples include subsidies for electricity from photovoltaic (PV) systems in the United Kingdom, Germany, Spain, California and other jurisdictions.

There are many criticisms of regulatory and, more particularly, technology subsidy policies. Detractors point out that governments are not necessarily good at picking technology 'winners' and that, once started, subsidies are difficult to phase out. Such policies can have high administrative costs and provide incentives to those who would have invested in the technology in any case. Poor subsidy design can have unintended consequences. For example, PV and wind subsidies initially increased demand and, therefore, technology costs. However, supporters point out that well-designed subsidisation policies have a place in the energy policy mix. Where market failure occurs, subsidies can be targeted to overcome barriers to the uptake of cost-effective interventions. For example, subsidised home energy audits can encourage the uptake of cost-effective measures such as loft insulation, draught proofing and light-emitting diode (LED) lighting. The subsidisation of emerging technologies with high cost-reduction potential can quickly unlock their market potential. The widespread subsidisation of wind energy, for example, means that it has now reached grid parity in many countries. Furthermore, the introduction of market-based incentives such as carbon taxes can be politically difficult such that subsidies represent a more feasible option for policy makers.

Where subsidies are used, it is important that they are well designed. Particularly, they should result in emissions mitigation, which is cost-effective for the taxpayer. The most suitable technologies and projects to achieve GHG reduction targets can be identified using the MAC of the technology, policy or project. The MAC is the cost of reducing GHG emissions by one unit by switching from the conventional (or baseline) technology to an alternative technology. Technologies with MACs less than a threshold value make them attractive to subsidise; such thresholds may be the unit penalty cost of failing to meet national emissions targets, a calculated social cost of carbon or an agreed market value for carbon (such as that prevailing on carbon markets). The subsidy must be designed to incentivise private investment at least cost to society. The minimum subsidy requirement is the breakeven cost to incentivise private investment which is established using financial parameters such as levelised cost of energy (LCOE) or net present value (NPV). Once designed, the overall benefit of the subsidy can be assessed using social cost–benefit analysis (CBA), which may incorporate a variety of welfare costs and benefits.

6.2 Energy policy overview

6.2.1 Policy instruments and targets

Current energy policies in most developed economies seek to achieve environmental sustainability and security of supply while maintaining cost competitiveness. In addition, policies should add to societal welfare while maintaining or enhancing equity. Within this context, many countries have explicit energy policy targets in the form of reductions in GHGs, enhanced end-use energy efficiency or renewable energy technology targets, or combinations of all three (see Table 6.1) and use a mixture of market-based, regulatory, RD&D (research, development and demonstration) and informational policy instruments to achieve their stated policy targets. A brief summary of selected targets and policy instruments is provided below.

Market-based instruments are policy instruments which put a cost on GHG emissions through the use of carbon prices and markets. Consequently, they provide an incentive to energy consumers to reduce the costs of their emissions by becoming more energy efficient, changing behaviour and/or switching to lower emissions energy supply technologies and processes. Such instruments include carbon taxes, markets for trading emission permits and other activity-based taxes directly linked to emissions. Carbon taxes are charged on the basis of the carbon content of fuels and therefore discriminate between the different technologies on the basis of fuel type and system efficiency. However, such taxes can increase inequality, for

Table 6.1 The main energy-related policy targets and instruments for selected countries.

Country/ region	Main targets			Main policy instruments
	Energy efficiency	**Emissions**	**Renewable technology**	
EU	20% improvement by 2020	20% reduction from 1990 levels by 2020	Increasing the share to 20% of primary energy use by 2020	A variety of policies implemented by member states including emissions trading, FITs, quotas and green certificates
China	Energy consumption per unit GDP decreasing by 16% by 2015	Endeavour to lower emissions per unit of GDP by 40–45% by 2020 compared to the 2005	to comprise 9.5% of energy production by 2015	FITs for multiple renewable energy sources
India	Increase energy efficiency by 20% from the period 2007–2008 to 2016–2017	Reduce energy emission intensity by 20% from the period 2007–2008 to 2016–2017	Add 30,000 MW of renewable energy capacity during 2012–2017.	
Australia		reducing by between 5% and 15–25% below 2000 levels by 2020	20% of electricity from renewable sources by 2020	Emissions trading mechanism

Source: IEA (2014).

example, by increasing fuel poverty in low income households, although well-designed fiscal measures can counteract these effects. Many jurisdictions have carbon taxes including British Columbia (Canada), Japan, Ireland, the United Kingdom, Norway, Finland, Sweden, Denmark, the Netherlands and Switzerland.

Another market-based approach is to create an emissions market by limiting GHG emissions from emitters such as power producers and industries. Permits or allowances are distributed to the participants who can either reduce their own emissions or purchase permits from industries with excess allowances through the market, depending on which is less costly to them. The European Emissions Trading Scheme (ETS) is an example of a carbon market, which involves more than 10,000 participants from 31 countries. Energy certificates such as the UK's Renewable Obligation Certificates (ROCs) and Italy's Certificati Verdi are other examples of market-based systems. A problem with market-based policy tools arises when too many emissions allocations are issued; this has led to the recent collapse in carbon prices on the ETS. A further market-based instrument specifically for the transport sector is road pricing, which directly charges for the use of transport infrastructure (such as roads) on the basis of environmental impact such as emissions, noise and congestion; the London Congestion Charge is one such example.

Governments also use *standards and regulations* to implement their energy policies and meet policy targets. These involve setting mandated technical or performance requirements for specific technologies. In the European Union (EU), for example, fleet average new passenger car CO_2 emissions targets are to be reduced from 140 g CO_2/km in 2008 to 130 g CO_2/km by 2015 and 95 g CO_2/km by 2021. Similarly, in the buildings sector national minimum energy performance requirements are set in building regulations to limit primary energy use, energy end use or CO_2 emissions, depending on the country; Figure 6.2 shows examples of minimum building energy standards stipulated in selected European countries. Regulation is a cruder instrument as compared to market-based approaches because it will incur costs in individual cases where there are no benefits. For example, a single occupant using only part of a large house may not benefit economically if required by regulation to insulate the entire dwelling. However, they remain an important tool in cases of clear market failure, such as in the case of split incentives between producer and user.

Subsidies are incentives that are provided for specific technologies, services or practices, which have been identified by policy makers to deliver energy policy objectives. The aim of subsidies is to incentivise technology uptake by reducing costs to the investor. Policy costs are distributed across societal groups such as taxpayers and energy consumers. They include capital subsidies, feed-in-tariffs (FITs) and favourable tax treatment. Strictly speaking, FITs are payments for every unit of electrical energy fed to the public electrical grid. However, this term is sometimes used to describe a subsidy for the certified production of a unit of thermal or electrical energy. Capital subsidies provide a once-off incentive to invest in the technology but, unlike FITs, they do not provide an ongoing incentive to use it. Subsidies are criticised for

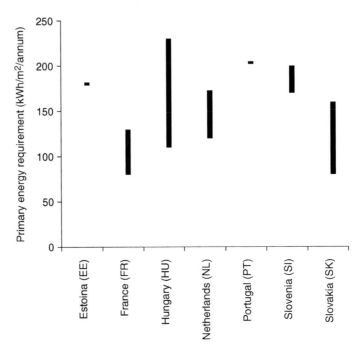

Figure 6.2 Regulated minimum primary energy requirement ranges for dwellings in various European countries (BPIE, 2011).

several reasons. They often produce unintended consequences: one example is the displacement of important agricultural production in developing countries with energy crops due to the EU and US biofuel subsidies. Subsidies may also increase the prices of the technologies they have been designed to promote because suppliers may initially increase costs. Furthermore, once introduced, they can be difficult to change or phase out for political reasons. Moreover, if they are too high or low they may encourage too much, or too little uptake of the desired technology. Nonetheless, subsidies can help overcome market failures and promote the development of immature technologies so that they can achieve the necessary economies of scale to compete with mature alternatives.

There are a variety of other energy policy instruments. These include *supports to research and development* at different stages of technology readiness to encourage the development and deployment of technologies, which will support policy objectives. *Informational policies* are important in overcoming information-related market failures.

6.2.2   Designing policy instruments

There is no single approach to the design of policy instruments to meet energy policy targets. Their choice depends on a wide variety of factors including the type of target (e.g. technology, end-use efficiency or emissions), government decision-making processes, decision criteria, data quality, the availability of economic models, the time frames involved and the capabilities of those involved. For example, multi-criteria decision analysis (MCDA) (see Chapter 5) can be used to shortlist a wide range of technology options to meet national emissions targets. Cost may not be the only or even most important criterion: approximate marginal abatement and policy costs based on initial assumptions might form inputs into the MCDA along with other factors such as employment, public acceptability and import substitution. Where emissions' reduction is the main policy objective, detailed MAC estimations may be used to choose between technologies to identify the set of least-cost alternatives for emissions abatement. Where renewable energy technology targets have been identified, analysis may focus on establishing least-cost solutions using LCOE. Finally, social cost–benefit analysis is used to establish the value of policies or public projects to society and identify their distributional effects. Data constraints will play an important role in whether and how accurately different methods are applied. For example, it is difficult to obtain information on the life cycle costs (LCC) of heating and cooling in an economy given the diverse nature of this energy end-use sector. Similarly, financial parameters such as discount rates for emerging technologies (e.g. ocean energy) may need to be approximated. Assumptions about future parameter values such as technology learning rates and energy price inflation are also subject to significant uncertainties.

There are many techniques which can be used for designing and assessing the welfare impacts of energy policies, some of which are summarised as follows and described in more detail over the remainder of this chapter.

- *Marginal Abatement Costing (MAC).* This can be used to estimate the cost of reducing emissions by switching to an alternative technology. It may be used to identify the least-cost options for meeting emissions reductions targets.
- *Net Present Value (NPV).* NPV can be used to establish the present value of the technology and therefore the subsidy (where negative) required to incentivise private investment. Where private NPVs are positive and the technology remains underutilised, then the policy maker should look for a market barrier, which is limiting uptake. NPV can be used in the design of 'top-up' or premium FITs and subsidies.
- *Levelised Cost of Energy (LCOE).* This can be used to establish the life cycle costs of producing one unit of energy from a technology. It can be used to identify what fixed FITs are required to incentivise technology investment in the private sector.
- *Social Cost–Benefit Analysis (CBA).* This can be used to check the value of supporting technology subsidies options to society and to establish their distributional effects for further compensatory policies where appropriate. All policies should be subject to this analysis.

- *Multi-Criteria Decision Analysis.* MCDA can be used to shortlist multiple technology options for further detailed analysis using MAC, NPV, LCOE and CBA.

6.3   Marginal abatement cost

A marginal cost is the change in total costs associated with the production of one additional unit. It may either be 'short run', which considers only the immediate incremental costs incurred, or 'long run', which considers all factors of production over the investment life cycle. For example, in a power station short-run marginal costs might only consider the additional costs of fuel and maintenance, whereas long-run marginal costs would consider investment, staff and other fixed costs. LCOE captures these long-term investment requirements and can be used as a measure of the long-run marginal cost of a technology. For the purposes of assessing policies which must be effective over decades, long-run marginal costs are normally used. Where energy forms are perfect substitutes, the cost of investment switching from one technology to another is the difference between their long-run marginal costs. Consequently, it is important to note that marginal costs are dependent on the reference case chosen.

Switching from one energy technology to another typically results in changes in GHG emissions. For example, replacing gas turbine electricity generation with offshore wind turbines will reduce emissions. However, this reduction in emissions often comes at an additional cost for the cleaner technology. The ratio of this additional cost to each unit of emissions avoided is the MAC.

The perfect substitute assumption is an important one which may not receive sufficient consideration in policy assessment. For example, in the case of replacing gas-fired generation with offshore wind, although both produce electricity, which is indistinguishable, the former is dispatchable (available for production when required) whereas the latter is intermittent (not necessarily available when required). A complete comparison would consider the costs of additional system services (e.g. storage, inertia, black start) required by wind. These costs are often ignored.

There are a number of different approaches to estimating the marginal abatement costs of energy supply and energy efficient technologies. These can be categorised as top-down economic models and bottom-up techno-economic models. Economic models that consider the effects of different energy supply and efficiency technologies on part or all of an economy are referred to as partial and general equilibrium models, respectively. The approach is appropriate as energy is an important input to all economic sectors and affects economic activity across the economy as a whole. The models take account of factors such as sectoral inter-relationships, price elasticities of demand, substitution to estimate economic costs and emissions at a regional or national level. This approach is beyond the scope of this book.

Although the bottom-up techno-economic approach does not consider the economy-wide impacts of switching technologies, it is a widely applied approach that allows individual technologies to be studied in detail (see, e.g. McKinsey and Company, 2009). The approach involves first estimating the life cycle GHG emissions from a technology, then assessing its LCOE and using these two parameters to establish the MAC. This approach is described in the remainder of this section.

6.3.1   Environmental life cycle assessment

Environmental life cycle assessment (LCA) is a systematic method for estimating all of the environmental impacts arising from the provision of a product or service over its life cycle, from raw material extraction to end-of-life. For an energy supply or energy-efficient system, it typically includes the entire supply chain including material extraction, manufacturing, construction, energy conversion, operation and maintenance and decommissioning. It includes all direct and indirect impacts. For example, a diesel generator will emit GHG emissions when burning fuel to create electricity; these are termed 'direct' emissions because they are a direct result of electricity generation. However, there are many other 'indirect' emissions resulting from electricity generation by the generator. The maintenance of the generator results in energy use and emissions since maintenance vehicles and office staff require energy for transportation and work. The generator itself had to be manufactured, steel had to be smelted and ore had to be extracted, all of which required energy and resulted in emissions. The diesel had to be extracted and refined. In fact, there is a large set of indirect activities which can be allocated to the electricity output from the generator, which become increasingly small the more indirect they are. However, taken together they can account for a significant fraction of the direct emissions. This is particularly true of energy-efficient or renewable energy technologies which have low or no direct emissions.

Environmental LCA is used to estimate many energy-related environmental impacts: sulphur dioxide from fossil fuel combustion results in defoliation and freshwater acidification; particulate diesel combustion matter has respiratory health effects; and nuclear power can result in radioactive contamination. However, we are most interested in the GHG emissions impacts of energy use because their mitigation is the main environmental focus of international energy policies. Although CO_2 is by far the most important energy-related GHG, there are other anthropogenic gases which contribute to global warming such as nitrous oxide (N_2O), methane (CH_4) and the 'F gases' including hydrofluorocarbons (HFCs) and perfluorocarbons (PFCs). The effects of these gases are normalised to a 'carbon dioxide equivalent' (CO_2-eq) using the global-warming potential (GWP) conversion factor (shown in Table 6.2) and

Table 6.2 Global warming potentials of the three most important energy-related greenhouse gases used in the calculation of carbon dioxide equivalent values.

Common Name	Chemical Formula	GWP			
		20-yr	100-yr	500-yr	SAR 100-yr
Carbon dioxide	CO_2	1	1	1	1
Methane	CH_4	72	25	7.6	21
Nitrous oxide	N_2O	289	298	153	310

The 100-year value provided in the IPCC's Second Assessment Report (SAR 100-yr) is typically used.
Source: IPCC (2007).

summed to give their combined impact. F-gases are omitted given the energy sector's very small contribution in this regard.

Example 6.1   Calculating CO_2-eq

A company operating a car fleet wishes to estimate the GHG emissions from its fleet in carbon dioxide equivalent. Over the year it consumed 81,864 litres of petrol. Given the CO_2, CH_4 and N_2O emission factors for petrol combustion of 2.32, 1.14×10^{-4} and 9.43×10^{-5} kg/l, respectively, calculate the CO_2-eq emissions for the year.

   Total emissions of each gas are first calculated by multiplying fuel consumption by the relevant emission factor. These are then multiplied by the global warming potential for each gas to give carbon dioxide equivalents. These are summed to give total CO_2-eq emissions for the year of 192,182 kg, as shown in Table 6.3.

Table 6.3 Table showing calculation of total CO_2-eq for Example 6.1.

Gas	Emission factor (kg/l)	Total emissions (kg)	GWP	CO_2-eq emissions (kg)
CO_2	2.32	189,648	1	189,648
CH_4	1.14×10^{-4}	9	25	234
N_2O	9.43×10^{-5}	8	298	2,300
			Total	192,182

14044 Environmental management – Life cycle assessment – Requirements and Guidelines provides an overall framework for undertaking LCA studies. It divides the process into four steps:

- defining the goal and scope;
- inventory analysis;
- impact assessment; and
- interpretation.

Goal and scope

The goal and scope definition phase defines the motivation, objectives, boundary and intended audience of the study. An important task in this phase is to define the 'functional unit', which is the unit that describes the function of the system being considered. For example, an emission study of dwellings may have functional units of emissions per metres squared of floor area, per occupant or per dwelling, depending on the exact motivation of the study. For many energy supply or efficient systems, functional units based on units of energy output are typically used.

Life cycle inventory analysis

Life cycle inventory analysis concerns quantifying the inputs and outputs to the system in question. A variety of approaches exist for estimating GHG emissions associated with different activities. These include process analysis, input–output (I–O) analysis and a variety of combinations of these techniques collectively referred to as 'hybrid analysis'; see Crawford (2008) for a comprehensive overview. All techniques involve obtaining the product of an activity level (e.g. kilometres driven or tonnes of steel used) and an emission intensity (e.g. emissions per kilometre or per tonne of steel) to give total emissions for a particular activity.

I–O analysis was first developed by Wassily Leontief in the 1930s and has had environmental applications since the 1970s. A national economy is divided into a number of sectors, each assumed to produce a single, uniform good or service. I–O tables, which are constructed from national accounts and trade balance data, show the extent to which a monetary output from one sector is an input to each of the other sectors. These can be combined with direct sectoral emissions data from environmental accounts to estimate the total sectoral emissions, including all upstream effects. The technique includes all upstream inputs, but assumes sectoral homogeneity, which can result in significant error at the level of an individual product or service. The emissions resulting from a product or service are estimated using I–O analysis as the product of the sectoral emissions' intensity and its monitory value according to the equation

$$LCE_x = \sum_{s=1}^{n} EI_s \times P_{x,s} \qquad (6.1)$$

where LCE_x are the life cycle emissions (LCEs) resulting from product or service x (kg CO_2-eq), EI_s is the I–O sectoral emission intensity of sector s (kg CO_2-eq/€), $P_{x,s}$ is the total cost over the life cycle component of product or service x which can be allocated to sector s (€) and n is the total number of economic sectors.

Sectoral emissions' intensities for EU-27 countries which were derived from I–O and emissions data for 2005 are given in Table 6.4.

Table 6.4 Sectoral emission intensities of consumption (g CO_2 - eq/€) for EU-27 countries for the year 2005 (authors' calculations).

Sector	NACE Code	Austria	Denmark	France	Germany	Italy	Netherlands	Norway	Portugal	Spain	Sweden	United Kingdom
Agriculture, forestry, fishing	1_5	1206.5	1369.8	1265.8	1038.8	828.9	947.0	1900.7	1137.4	990.9	1061.8	1534.1
Mining and quarrying	10_14	198.1	214.2	157.2	437.0	245.3	254.5	197.9	443.8	385.0	128.0	332.0
Manufacture of food products; beverages and tobacco	15-16	401.1	696.6	511.6	461.8	499.6	463.4	709.6	650.2	573.2	363.9	434.6
Manufacturing of textile and leather products	17-19	154.9	224.3	135.1	273.7	306.8	202.3	164.7	357.9	294.7	128.6	190.6
Manufacture of wood and wood products	20	370.7	250.9	338.7	356.5	269.5	198.1	428.8	682.9	353.6	343.2	315.5
Manufacture of pulp, paper and paper products; publishing and printing	21-22	347.3	183.9	181.7	273.3	354.0	190.3	199.6	528.3	341.3	275.7	232.1
Manufacture of coke, refined petroleum products and nuclear fuel	23	346.3	288.4	367.1	548.7	572.8	453.8	n.a.	755.4	733.2	256.6	538.5
Manufacture of chemicals, chemical products and man-made fibres	24	252.5	189.4	253.4	365.4	323.6	482.6	323.2	747.0	415.8	149.8	323.3
Manufacture of rubber and plastic products	25	164.4	189.8	177.7	263.5	672.9	278.2	146.6	479.1	315.2	117.5	323.6
Manufacture of other non-metallic mineral products	26	677.1	837.0	742.1	958.2	1239.4	410.4	528.0	1655.8	1513.9	729.4	762.1
Manufacture of basic metals	27	805.1	265.6	438.7	680.9	557.9	482.0	441.9	457.9	542.9	334.0	881.1
Manufacture of fabricated metal products, except machinery and equipment	28	236.4	174.6	179.2	290.4	249.1	206.3	163.7	339.5	324.5	128.9	314.8

Manufacture of machinery and equipment n.e.c.	29	160.2	147.5	129.0	185.2	251.7	140.5	121.0	291.5	246.8	100.7	249.9
Manufacture of office machinery and computers	30	99.0	106.3	105.4	130.0	199.9	172.8	71.4	214.2	223.3	85.0	137.3
Manufacture of electrical machinery and apparatus n.e.c.	31	183.2	179.9	139.4	181.6	274.7	213.9	153.2	315.9	318.6	79.1	219.0
Manufacture of radio, television and communication equipment and apparatus	32	130.3	127.1	111.5	150.3	178.1	175.8	76.6	267.3	262.3	n.a.	163.9
Manufacture of medical, precision and optical instruments, watches and clocks	33	95.7	101.9	106.8	142.9	184.8	118.5	82.7	238.9	214.6	73.5	134.8
Manufacture of motor vehicles, trailers and semi-trailers	34	148.0	168.8	161.8	227.5	293.9	161.9	160.4	314.3	298.3	112.3	256.7
Manufacture of other transport equipment	35	141.9	173.4	118.2	196.4	244.4	144.9	117.0	222.2	245.1	92.6	202.7
Manufacturing n.e.c.	36-37	176.0	195.5	233.9	209.8	256.2	153.0	150.6	368.2	258.6	152.4	257.2
Electricity, gas and water supply	40-41	1156.6	2358.2	660.9	3218.0	1943.2	2165.2	74.6	2926.7	2222.4	613.2	2417.9
Construction	45	207.9	220.7	150.0	213.2	262.3	167.4	159.2	520.1	303.3	193.5	175.6
Wholesale and retail trade, repair of motor vehicles, etc.	50-52	111.9	208.9	92.9	155.6	199.1	122.5	87.4	245.2	199.6	91.6	159.6
Hotels and restaurants	55	139.5	282.9	197.2	217.1	249.1	246.9	149.9	350.3	195.6	127.8	158.8
Transport, storage and communication	60-63	332.1	1004.4	284.3	422.6	349.2	566.4	499.6	705.7	512.7	352.2	552.4

(continued overleaf)

Table 6.4 (*continued*)

Sector	NACE Code	Austria	Denmark	France	Germany	Italy	Netherlands	Norway	Portugal	Spain	Sweden	United Kingdom
Post and telecommunications	64	74.7	119.8	51.4	118.2	132.6	72.6	92.1	132.2	171.1	74.2	110.9
Financial intermediation	65-67	52.6	37.9	46.5	63.4	54.6	44.9	30.2	74.2	57.9	28.0	78.0
Real estate activities	70	75.7	43.7	19.3	34.6	30.6	43.3	46.1	72.9	60.0	83.1	35.0
Renting of machinery and equipment without operator, etc.	71	54.0	111.3	181.5	23.3	224.9	221.3	84.7	142.2	147.2	91.9	127.2
Computer and related activities	72	61.0	91.8	53.9	38.8	113.0	67.3	43.2	139.2	72.6	56.7	54.8
Research and development	73	99.2	102.7	109.3	110.8	89.9	139.4	66.9	118.9	131.5	56.8	84.4
Other business activities	74	81.6	88.5	55.1	58.8	126.2	96.8	52.8	186.5	142.9	0.0	64.8
Public administration and defence; compulsory social security	75	70.9	101.4	81.5	120.2	99.8	140.6	67.1	250.5	125.9	74.0	141.9
Education	80	71.0	84.5	80.9	106.8	36.6	81.7	36.0	82.1	66.1	53.5	88.1
Health and social work	85	86.0	86.0	63.8	117.8	105.5	102.7	50.4	307.8	118.6	39.5	114.8
Other community, social, personal service activities	90-93	258.8	190.8	280.2	231.3	423.3	590.7	171.2	880.4	335.8	184.5	245.8
Activities of households as employers of domestic staff	95	0.0	0.0	0.0	0.0	0.0	0.0	0.0	0.0	0.0	0.0	22.6

Abbreviation: n.e.c., not elsewhere classified.
'NACE' stands for 'Nomenclature generale des Activites economiques dans les Communautes europeennes' and is the industrial sector classification system used in the European Union.

Example 6.2   Input–Output Analysis

A 2.5-MW wind turbine is erected in Denmark at a cost of €3.75m. If annual maintenance costs amount to €175,000 estimate LCEs using I–O analysis assuming a 20-year lifespan.

The sectors which are deemed to be most representative of the activity on which money has been spent are first identified. The construction of the wind turbine is attributed to the 'construction' sector and maintenance to the 'manufacture of machinery and equipment' sector. The relevant sectoral emissions intensities are chose in from Table 6.4. The products of expenditures and sectoral emissions intensities are then summed over the lifespan of the project using Equation 6.1, as shown in Table 6.5.

Table 6.5   Results for Equation 6.2.

Cost category	Cost (€m)	Appropriate sector	Emission factor (gCO$_2$-eq/€)	Annual emissions (tCO$_2$-eq/annum)	Years of emissions (yr)	Life cycle emissions (tCO$_2$-eq)
Wind turbine	3.750	Construction	220.7	828	1	828
Maintenance	0.175	Manufacture of machinery and equipment	147.5	26	30	774
					Total	1602

Process analysis first involves analysing the supply chain upstream of a product or service and identifying the most important direct and indirect activities. These are quantified in appropriate units (e.g. construction materials used in tonnes, freight transport in tonne-kilometres or liquid fuels in litres). Activity data are then combined with relevant emissions intensities or factors and summed to give total emissions for a particular product or service, as shown in Equation 6.2.

$$\text{LCE}_x = \sum_{a=1}^{n} \text{EI}_{a,i} \times Q_{x,a} \tag{6.2}$$

where LCE_x denotes the LCEs resulting from product or service x (kg CO$_2$-eq), EI_a is the process emissions intensity of activity a (kg CO$_2$-eq per kg, l, kWh, km or other measure), $Q_{x,a}$ is the quantity of activity a (kg, l, kWh, km or other measure) used to produce the product or service over its life cycle and n is the total number of materials used. Selected emission intensities for different materials are shown in Table 6.4 and some databases of emissions intensities are shown in Table 6.6.

Table 6.6   Sources of data on emission intensities.

Database	Provider	Web site
Inventory of Carbon and Energy	University of Bath, UK	www.circularecology.com/ice-database.html
US Lifecycle Inventory Database	National Renewable Energy Laboratory, USA	www.lcacommons.gov/nrel/search
European reference Life Cycle Database	Joint Research Commission, EU	eplca.jrc.ec.europa.eu/ELCD3/index.xhtml
Open LCA	GreenDelta, Germany	www.openlca.org/openlca
CMLCA	Leiden University, Belgium	www.cmlca.eu/
CPM LCA database	Swedish Life Cycle Center (CPM), Sweden	cpmdatabase.cpm.chalmers.se/

Although process-based emission intensities are accurate for the extent of the supply chain analysed, it is ultimately and often unknowingly necessary to truncate the supply chain and disregard a very large number of activities which individually have a relatively small effect, but collectively can represent a significant error.

Example 6.3   Process Analysis

Calculate the LCEs of a wind turbine which has a steel tower/nacelle, composite epoxy blade assembly and reinforced concrete foundations weighing 127, 36 and 480 t, respectively. Maintenance is undertaken using a diesel van travelling 20,800 km/annum. The project has a 20-year lifespan. Decommissioning emissions are ignored because these are small. Only limited construction materials are included for simplicity; only maintenance-related fuel is included for the same reason.

Solution: Applying Equation 6.2 to the quantities of materials given and the relevant process emission factors from Table 6.7 we get the results summarised in Table 6.8.

Table 6.7   Emission intensities for selected materials.

Material	Energy intensity (MJ/kg)	Carbon intensities	
		$(kgCO_2/kg)$	$(kgCO_2\text{-eq/kg})$
Plate-UK (EU) average recycled content	25.10	1.55	1.66
Epoxide resin	137.00	5.70	n.a.
Reinforced RC 25/30 MPa (with 100 kg/m³ concrete)	1.82	0.18	0.19

Abbreviation: n.a., not applicable.
Source: Hammond and Jones (2011).

Table 6.8   Results for Example 6.3.

Element	Quantity	Emission factor	Notes on emission factors	Annual emissions (tCO_2-eq/annum)	Years of emissions (yr)	Life cycle emissions (tCO_2-eq)
Tower/nacelle	127,000 kg	1.66 $kgCO_2$-eq/kg[a]	Plate-UK (EU) average recycled content	211	1	211
Blades assembly	36,000 kg	5.70 $kgCO_2$-eq/kg[a]	Epoxide resin – CO_2 emissions	205	1	205
Foundations	480,000 kg	0.19 $kgCO_2$-eq/kg[a]	25/30 MPa Conc plus reinforcing	91	1	91
Maintenance	20,800 km	283.00 gCO_2-eq/km[b]	Table 4.5: estimated emission factors for diesel light duty vehicles	5.9	20	118
				Total		625

[a] *Source*: Hammond and Jones (2011).
[b] *Source*: UK National Atmospheric Emissions Inventory (http://naei.defra.gov.uk/).

Process and I–O emissions accounting techniques can be combined to benefit from their respective strengths while minimising the effects of their weaknesses. Suh and Huppes (2005) describe three approaches to hybrid analysis: tiered, I–O-based and integrated; the choice of technique depending on factors including data availability, available analytical tools, expertise, time constraints, system boundaries and required accuracy. Similarly, Crawford (2008) identifies two hybrid methods: process-based and I–O-based, the former (see Bullard et al., 1978) can significantly underestimate aspects (such as pollutant emissions) for individual products and systems, whereas the latter is systemically complete. The simplest, but, by no means, the most comprehensive way to combine these techniques is to use process analysis, where material quantities are available, and to add I–O analysis to the values of the balance of materials and activities. Other, more complex hybrid approaches are beyond the scope of this book. In practice, however, process analysis is normally used for commercial LCA studies.

In summary, I–O analysis is complete when applied to sectors where the product/service can be treated homogenously; this has obvious applications in policy analysis. However, when individual products/services are analysed, process analysis can be applied where system boundaries are limited and data applicable; for more complete boundaries, an appropriate hybrid approach should be adopted.

Example 6.4   Process-Based Hybrid Analysis

A 2-GW gas-fired power station is built in the United Kingdom in 2005 at a cost of €1bn, comprising €400m mechanical plant, €300m electrical equipment and €300m civil works. Operation and maintenance costs 2 €/MWh per unit of electricity generated and gas costs 30 €/MWh. Assuming a 30-year lifespan and a capacity factor of 0.5, estimate its LCEs. Plant efficiency is 50%. Ignore decommissioning costs.

The overall approach here involves the preferred use of process analysis where materials quantities exist and the secondary use of I–O analysis where only prices exist, taking care not to double the count. Therefore, fuel (gas) quantities used are first estimated and then multiplied by their corresponding emission factor. I–O analysis is then applied to monetary quantities using the most appropriate sectoral emission intensity (see Table 6.4).

The annual electricity output is the product of its rated power, the number of hours in the year and capacity factor:

$$2 \text{ GW} \times 8760 \text{ h} \times 0.5 = 8760 \text{ GWh/annum}$$

Annual maintenance costs are the product of annual electricity output and the maintenance cost of 2000 €/GWh:

$$8760 \times 2000 = €17.52\text{m/annum}$$

Annual gas consumption is electricity output divided by efficiency (Equation 2.3):

$$\frac{8760}{0.5} = 17{,}520 \text{ GWh/annum}$$

Taking an emission factor for natural gas of 204.7 tCO_2-eq/GWh and a lifespan of 30 years, Equation 6.2 gives

$$\text{LCE}_x = \sum_1^{30} 204.7 \times 17{,}520 = 107{,}590 \text{ kt } CO_2\text{-eq}$$

The expenditures on mechanical plant, electrical equipment, civil works and maintenance are assigned to the closest economic sectors listed in Table 6.4. The resulting emission factors are multiplied by monetary values and summed over the appropriate period according to Equation 6.1; plant installation emissions occur only once so are assigned 1 year of emissions whereas maintenance emissions occur over 30 years. Results are summarised in Table 6.9.

Finally, total LCE is the sum of the total process and the I–O emissions:

$$\text{LCE} = 107{,}590 + 350 = 107{,}940 \text{ kt } CO_2\text{-eq}$$

Table 6.9 I–O analysis of the gas-fired power station.

Cost category	Cost (€m)	Appropriate sector	Emission factor (gCO_2-eq/€)	Annual emissions (tCO_2-eq/annum)	Years of emissions (yr)	Life cycle emissions (tCO_2-eq)
Mechanical plant	400.00	Manufacture of machinery and equipment	249.9	99,960	1	99,960
Electrical equipment	300.00	Manufacture of electrical machinery and apparatus	219.0	65,700	1	65,700
Civil works	300.00	Construction	175.6	52,680	1	52,680
Maintenance	17.52	Manufacture of machinery and equipment	249.9	4,378	30	131,347
					Total	349,687

Impact assessment and interpretation

The remaining two phases in the LCA process are impact assessment and interpretation. The former involves evaluating the significance of the environmental emissions on the environment. However, this is not necessary in the case of GHG emissions, because they are already expressed in terms of their potential contributions to global warming. Furthermore, policy targets are normally expressed in terms of GHG emissions, not their impacts. Interpretation involves analysing the inventory and impact phases to identify important issues, sensitivities of inputs parameters, model verification and conclusions and recommendations.

6.3.2 Estimating marginal abatement costs

The MAC of switching from one technology to an alternative is given by

$$MAC_x = \frac{LCOE_x - LCOE_{dt}}{EI_{dt} - EI_x} \tag{6.3}$$

where MAC_x is the marginal CO_2-eq abatement cost of energy production of technology x compared to production using the alternative technology (€/tCO_2-eq), $LCOE_x$ is the levelised cost of producing one unit of energy using technology x (€/kWh), $LCOE_{dt}$ is the levelised cost of producing one unit of energy using the displaced technology dt (€/kWh), EI_{dt} is the emission intensity of the displaced technology dt (tCO_2-eq/kWh) and EI_x is the emissions intensity of technology x (tCO_2-eq/kWh).

The emissions intensity of a technology is given by

$$EI_x = \frac{LCE_x}{\sum\limits_{n=1}^{N} E_{x,n}} \tag{6.4}$$

where EI_x is the emission intensity of technology x, LCE_x denotes the LCEs of technology x, $E_{x,n}$ is the annual energy sold for technology x and N is the lifespan of the technology.

Example 6.5   Emissions Intensities and Marginal Abatement Cost

Estimate the emission intensity for Example 6.4. The new gas-fired power plant is competing with coal-fired technology. If the emission intensity of the displaced coal technology is 820 g CO_2-eq/kWh and the levelised costs of the gas and coal technologies are 50 and 40 €/MWh, respectively, calculate the MAC of switching to the gas-fired technology.

First, we calculate the emissions intensity of the gas-fired plant. From the solution to Example 6.4, we know that the LCEs are 107,940 kt CO_2-eq and that annual electricity output is 8760 GWh. Using Equation 6.4 we get

$$EI_{gas\ plant} = \frac{107,940}{\sum\limits_{1}^{30} 8760} = 616\ tCO_2\text{-eq/GWh} = 0.616\ tCO_2\text{-eq/MWh}$$

We then take the emissions intensities for the coal- and gas-fired technologies of 0.820 and 0.616 tCO_2-eq/MWh, respectively, along with the levelised costs of 40 and 50 €/MWh and insert these into Equation 6.3 to get (where coal is the displaced technology):

$$MAC_{gas\ plant} = \frac{LCOE_{gas\ plant} - LCOE_{coal\ plant}}{EI_{coal\ plant} - EI_{gas\ plant}} = \frac{50 - 40}{0.820 - 0.616} = 49\text{€/}tCO_2\text{-eq}$$

In the specific case where the energy outputs from two technologies are identical, MAC is also given by

$$MAC_x = \frac{LCC_x - LCC_{dt}}{LCE_{dt} - LCE_x} \tag{6.5}$$

where LCC_x is the discounted LCC of technology x (€), LCC_{dt} is the discounted LCC of the displaced technology dt (€), LCE_{dt} is the discounted life cycle emissions of the displaced technology dt (tCO_2) and LCE_x is the discounted LCEs of technology x (tCO_2-eq).

Example 6.6   Marginal Abatement Costs for Technologies with Identical Energy Outputs

A large industry has recently installed a new coal-fired furnace. However, a carbon tax of 30 €/ton has just been introduced and it is considering replacing it with a lower carbon technology. You are required to estimate the MAC of replacing the coal-fired furnace with a gas-fired furnace, both of which have 20-year lifespan. The investment costs, operating costs and annual CO_2-eq emissions for both technologies are given in Table 6.10. A discount rate of 6% is applied.

Table 6.10   Costs and emissions for the coal- and gas-fired furnaces.

Technology	Capital costs (€m)	Annual operating (€m)	Construction emissions (tCO$_2$-eq)	Annual emissions (tCO$_2$-eq)
Coal-fired boiler	0	12.129	0	235,846
Gas-fired boiler	6	12.367	3100	103,059

The coal-fired boiler is given a cost of zero because it has just been purchased and therefore is a sunk cost. Adapting Equation 4.22 for LCCs in Chapter 4 the LCC for technology x can be written as

$$LCC_x = C_{c,0} + C_{no} CDF_{N,d} \qquad (6.6)$$

where $C_{c,0}$ is the investment cost in year zero; C_{no} is the real recurrent annual operating cost; and $CDF_{N,d}$ is the cumulative discount factor for a N-year lifespan and real discount rate d. Using cumulative discount factor of 9.637 for a real discount rate of 6% and a 15-year lifespan we get

$$LCC_{coal} = 0 + 12.129 \times 9.712 = €117.80 \text{ m}$$

$$LCC_{gas} = 6 + 12.367 \times 9.712 = €126.11 m$$

Discounted LCEs for technology $x(LCE_x)$ are then calculated in a similar way using the equation

$$LCE_x = E_{c,0} + E_{no} CDF_{N,d}$$

where $E_{c,0}$ are the construction emissions in year zero, E_{no} are the recurrent annual emissions and $CDF_{N,d}$ is the cumulative discount factor for a N-year lifespan and real discount rate d. Using this equation, we get

$$LCE_{coal} = 0 + 235,846 \times 9.712 = 2,290,538 \text{ tCO}_2\text{-eq}$$

$$LCE_{gas} = 3100 + 103,059 \times 9.712 = 1,000,907 \text{ tCO}_2\text{-eq}$$

The marginal abatement cost is then calculated using Equation 6.5, where coal is the displaced technology:

$$\text{MAC}_x = \frac{126{,}109{,}000 - 117{,}800{,}000}{2{,}290{,}538 - 1{,}000{,}907} = 6.50 \text{ €/tCO}_2\text{-eq}$$

As this is significantly lower than the new carbon tax rate, the investment in the new gas furnace would represent good value for money.

MAC is not the only factor in identifying technologies which are most suitable for reaching emissions targets. A technology may have a very low MAC, but it may not be possible to deploy sufficient quantities at a regional level to justify the transaction costs required. For example, in a high density urban area, domestic-scale PV may have such a limited deployment potential that it would be better to use policy-making resources elsewhere. Therefore, MAC should be accompanied by a market assessment to estimate the deployment potential of the relevant technology.

The choice of the displaced technology is centrally important to the MAC obtained as this determines both the reference levelised cost and the emission intensity. The approach to this decision depends on the study being undertaken, but a number of factors should be considered. The main decision is whether the technology will displace a marginal or average unit of energy generated. For example, small deployments of wind turbines could justifiably use the average unit of grid electricity as the reference technology where they will displace all forms of electricity generation equally. However, the large-scale deployment of wind power may displace investment in new gas-turbine or other specific technology, which would then be a more appropriate choice of displaced technology.

Technology learning should also be considered when using MACs in policy assessment. Policy implementation can take years, by which time the economic and environmental performances of technologies could have changed significantly, particularly so for emerging renewable technologies. Therefore, MACs should anticipate the likely performance of technology options at the point in time when policies are likely to be implemented. This can be achieved using technology learning rates together with projected market deployment.

Normally MAC curves such as those in Figure 6.3 are assessed and presented individually. However, where a portfolio of abatement technology options is being considered, MACs must be estimated together rather than individually in order to account for technology interactions.

6.4 Subsidy design

6.4.1 Types of energy subsidies

Subsidies are payments to producers which increase the supply of products or services by lowering their costs, so that they become more affordable in

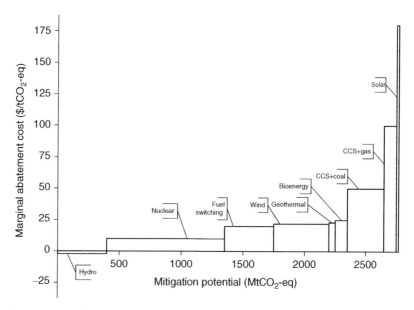

Figure 6.3   Indicative marginal abatement costs and potentials for selected electricity generation technologies in Organisation for Economic Co-operation and Development (OECD) countries (various data sources).

the market. Energy subsidies are typically funded by the taxpayer or energy consumers and their use is widespread in the sector. Global subsidies of fossil fuels and renewables were approximately $544bn and $101bn, respectively, in 2012 (IEA, 2013). Many types of subsidies are used internationally to support energy in the domestic, transport and industrial sectors. These include supports to production (e.g. FITs), employment (e.g. special tax-treatment of salaries) and consumption (e.g. price regulation).

Production subsidies in the energy sector can be divided into capital subsidies and FITs. The former may either be capital grants or favourable tax treatment such as accelerated capital allowances. Capital subsidies for technologies incentivise capital investment, not their operation and are therefore more suited to measures which have no operational costs, such as insulation upgrades. Where capital subsidies are used to support technologies that have higher operational costs than competing alternatives, a project may be abandoned during the operational phase, thus wasting public money. FITs, on the other hand, are popular because they incentivise the ongoing operation of energy-efficient and renewable energy projects. They are typically long-term agreements with energy producers which guarantee them either a base price or a 'top-up' on market prices for energy generated. Their intention is normally to accelerate the uptake of preferred technologies until they become commercially viable, at which stage they are removed.

In this section, we focus primarily on the design of FITs due to their international popularity in incentivising renewable energy and energy-efficient investments. However, the approaches described can be applied to the design of other subsidy types.

6.4.2 Calculating feed-in-tariffs

FITs can be classified as 'fixed' or 'premium'. Fixed tariffs provide a guaranteed price to the producer for the energy produced, whereas premium tariffs offer a 'top-up' over and above the market price for energy. Figure 6.4 shows the market price for energy, premium and fixed FITs over time and illustrates the relationship between market prices and premium tariffs. Fixed FITs are based on the LCOE of a technology – that is, the minimum lifecycle cost per unit of energy which will provide an NPV of zero to the private investor at their discount rate, thus incentivising them to invest. Premium FITs are designed to provide the additional revenues required to make a technology viable after it has earned the market price for energy. In this case, the unsubsidised project returns a negative NPV, which must be augmented to increase the NPV to zero.

FITs are commonly set at a level that incentivises the last megawatt of technology investment necessary to meet the policy target. Take, for example, onshore wind. As this technology is deployed, in the short-term its marginal cost tends to increase as sites with the best wind resource are exploited first. Lower resource sites have lower capacity factors and higher unit costs; therefore, a FIT must be set at a level which incentivises the development of sites with wind resources that give the desired cumulative capacity. This concept is illustrated in Figure 6.5, which indicates the inverse relationship between site full-load hours and the levelised cost of generating wind energy. Therefore, the choice of the representative project for estimating LCOEs and

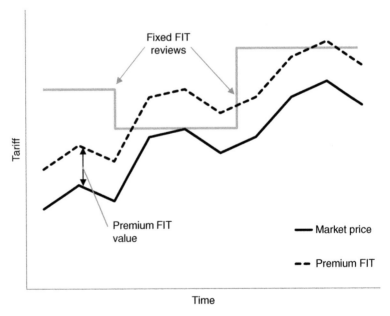

Figure 6.4 The relationship between energy market prices, premium and fixed feed-in-tariffs.

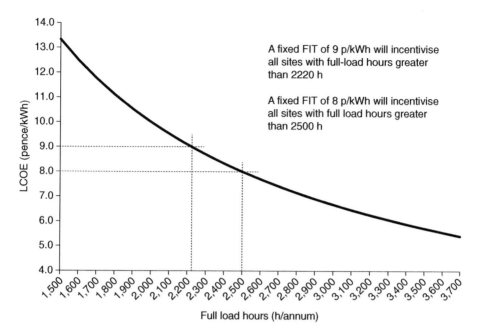

Figure 6.5   The effect of site wind resource characteristics (measured as full-load hours) on levelised cost of wind energy. Different FIT levels will incentivise the development of sites with different wind resource thresholds and, therefore, different deployment capacities.

NPVs is important for a well-designed FIT. This is normally established by a systematic expert study of the available resource and technology costs. In the case of onshore wind, for example, it involves mapping the deployment potential of different sites taking into account important cost factors such as wind resource, grid connection costs and proximity to population centres. It is an inexact science which relies on uncertain data, thus resulting in uncertainty about the suitability of the FIT in achieving its intended targets. Therefore, a complete approach to FIT design must involve the application of uncertainty analysis as described in Chapter 3 (sensitivity, scenario and Monte Carlo analyses).

An important component of a successful FIT policy is regular review. Technology learning is a significant medium-to-long-term factor in ongoing unit cost reductions for some renewable energy technologies as they are rapidly deployed in international markets. This puts downward pressures on LCOEs, which may necessitate reducing FITs for individual technologies over time if value-for-money to society is to be maintained. A countervailing effect however, is that as a technology is deployed and the most attractive projects are used up, the marginal costs increases and upward pressure is put on FITs if further deployment is to be achieved. In addition, fluctuations in fossil fuel prices will result in the need for readjustments to premium FITs. The term 'flexible' FIT is used to describe the regular review of tariffs to meet changes in these, and other technology input cost and profitability factors.

The fixed FIT for a representative project is calculated as its LCOE (Equation 6.7). A premium technology FIT can be calculated as the levelised energy subsidy required to increase the NPV of the project to zero; this assumes a negative NPV before subsidies are applied (Equation 6.8). A capital subsidy (CS) is simply given by the absolute value of the negative NPV of the representative project (Equation 6.9). In all cases, private discount rates which are representative of the industry are used since incentivising private investment is being considered.

$$FIT_{f,x} = LCOE_x \qquad (6.7)$$

where $FIT_{f,x}$ is the fixed FIT for technology x (€/MWh) and $LCOE_x$ is the levelised cost of the representative project for technology x (€/MWh).

$$FIT_{p,x} = \frac{-NPV_x}{\sum_{n=1}^{N} [E_n/(1+d)^n]}, \quad NPV_x \leq 0 \qquad (6.8)$$

where $FIT_{p,x}$ is the premium FIT for technology x (€/MWh), NPV_x is the NPV of the representative project for technology x (€), N is the project lifespan (in years), E_n is the energy sold in year n (in MWh) and d is the discount rate representative of the industry.

$$CS_x = -NPV_x, \quad NPV_x \leq 0 \qquad (6.9)$$

Example 6.7   Fixed FIT Design Using LCOE

A government energy agency wishes to set a fixed FIT for new onshore wind farms. Following an analysis of various data sources from industry, government departments, filed company accounts and the media as well as an assessment of wind resources on remaining sites, it establishes the following typical per-megawatt costs for new wind farms:

Capital cost	€1.159 m/MW
Cost of sales	€0.113 m/MW/annum
Weighted average cost of capital (WACC)	6% (real)
Capacity factor	34%
Lifespan	20 years
Decomissioning costs	Negligible

For simplicity, the effects of tax and depreciation are ignored in this example.

LCCs are established using Equation 6.6 with a cumulative discount factor of 11.470 (6% discount rate and 20-year lifespan):

$$LCC_{1MW\ wind} = 1.159 + 0.113 \times 11.470 = €2.455\ m$$

The annual electrical output (E_n) per megawatt is the product of the rated power (1 MW), capacity factor and number of hours in the year:

$$E_n = 1 \times 0.34 \times 8760 = 2978.4 \text{ MWh}$$

Discounted energy output is calculated using Equation 4.31:

$$\text{LCOE} = \frac{\text{LCC}}{\sum_{n=1}^{N}[E_n/(1+d)^n]}$$

$$\text{LCOE}_{1\text{MW wind}} = \frac{2.455}{\sum_{n=1}^{20}[2{,}978.4/(1+0.06)^n]} = 71.9 \text{ €/MWh}$$

Applying Equation 6.7 the FIT for new wind farms is set at 71.9 €/MWh. The energy agency should continue to review this FIT from time to time in order to capture changes in

- capital and operational costs;
- costs of finance;
- wind turbine technology, which tends to improve capacity factors over time; and
- the availability of high wind-resource sites, which tends to reduce capacity factors over time.

Example 6.8   Capital Subsidy Design Using NPV

A government agency wishes to establish what capital subsidy would be needed to incentivise new wind farm development given a future average spot market price estimate of 53 €/MWh and using the data in Example 6.7.

Revenues per MW are the product of the spot market price and the annual output of 2978.4 MWh, giving 157,855 €/MW/annum. Net operating cash flows are the difference between these revenues and cost of sales of 113,000 €/MW/annum giving 44,855 €/MW/annum. The NPV is then calculated using Equation 4.22:

$$\text{NPV} = -1{,}159{,}000 + 44{,}855 \times 11.470 = -644{,}511 \text{ €/MW}$$

The CS is determined by applying Equation 6.9:

$$\text{CS}_{1\text{MW wind}} = 644{,}511 \text{ €/MW}$$

Example 6.9 Premium FIT Design Using NPV

The agency further wants to establish the premium FIT which should apply to new onshore wind investment.

Using Equation 6.8 and inserting the NPV from Example 6.8, the annual electrical output (E_n) per MW from Example 6.7 at the 6% discount rate we get:

$$\text{FIT}(p)_{1\text{MW onshore}} = \frac{-(-644,511)}{\displaystyle\sum_{n=1}^{20}[2,978.4/(1+0.06)^n]} = 18.9 \ \text{€/MWh}$$

6.5 Social cost–benefit analysis

Social CBA is an economic method for estimating the net effects of the costs and benefits of projects or policies on societal welfare. It includes all private financial costs and benefits (described in Chapter 4) but extends the analysis to cover external costs such as impacts on the environment, employment and other economic sectors, which are not normally considered in private financial assessments. The decision criterion is that the project or policy represents a potential 'Pareto improvement', where at least one individual is better off without making anyone worse off. In practice projects where the overall benefits outweigh costs will have both winners and losers; however, a potential Pareto improvement occurs in those projects where the beneficiaries can compensate the losers, referred to as the Kaldor–Hicks criterion. Costs and benefits are discounted and expressed as an NPV or cost–benefit ratio (CBR) of (somewhat counter-intuitively given the parameter name) benefits to costs. Where an NPV is greater than or equal to zero, or where the CBR is greater than or equal to 1, the project is desirable, subject to certain conditions, which are discussed later in this section.

The cost of a project is its opportunity cost, which is the value of the best alternative foregone as a result of diverting resources into the project or policy; benefits are the valuable outputs from the project. The steps in a CBA can be classified as

1. define the objective and identify base case;
2. identify costs and benefits;
3. value costs and benefits;
4. discount costs and benefits;
5. interpret results;
6. assess who bears the benefits and costs;
7. sensitivity analysis; and
8. make decision.

6.5.1   Define the objective and identify base case

Before embarking on the CBA project, the project must be clearly defined and the objectives and scope must first be clearly identified. For example, it is often best to disaggregate policies as much as possible, so that those that may be unattractive are separable from those that are desirable. The policies being considered must be carefully chosen (based on MCDA or MAC analysis, e.g.), so that they represent the best available options. Geographical boundaries must be set. For example, for national policies, only GHGs that arise in the jurisdiction might be considered; 'imported' emissions may be discounted, even if these represent a significant portion of the project life cycle. Often national boundaries delimit the study, but locally funded projects may necessitate a more limited boundary. Conversely, supranational projects, such as transmission interconnectors, are assessed internationally.

A CBA cannot be undertaken in isolation; it must have a benchmark against which the project is assessed. The choice of this alternative has an important bearing on the final result and must be a fair representation of the world which would evolve without the project over its lifetime. Care must be taken to have realistic alternatives rather than very pessimistic ones, which will exaggerate the project's attractiveness. This step requires consultations with all relevant stakeholders such as policy makers, specialists, investors and affected communities. The choice and parameterisation of this alternative scenario can be very difficult, given that some energy projects can have very long time horizons over which many societal changes will take place. Remember that 'do nothing' may not be a valid option.

6.5.2   Identify costs and benefits

The next step is to identify the relevant costs and benefits of the project or policy compared to the reference project. It is important to identify all possible costs and benefits at this stage so that they are represented in the study – they can always be excluded later on. Costs are the opportunity costs of societal resources used in the project or policy and include land, labour and capital as well as environmental and social costs. They comprise both private costs such as those identified in Chapter 4 as well as public, external costs. For example, a large hydroelectric project requires land, labour and materials in its construction and operation; in addition, it will have significant impacts on aquatic life and will require the displacement of societies in the flooded areas. Similarly, benefits are any gains to society resulting from the project compared to the reference case. In the case of the hydroelectric dam, they may be energy, water and flood alleviation. Taxes are normally ignored because they can be treated as income transfers between the private sector and society, but because CBA measures societal welfare (which includes private investors), these transfers do not normally add to or subtract from the value of the project.

Table 6.11   Examples of costs and benefits for a renewable or energy efficient project or policy.

Costs	Benefits
Labour	Energy saved or produced
Investment	Health benefits
Maintenance and operation	Security of supply
Land values	Avoided fuel imports
Subsidies	
Health costs	
Negative externalities	
Climate change	

The foregoing are examples of 'direct' costs and benefits which are immediately attributable to the project. However, the identification and quantification of costs and benefits is more complex where the project or policy affects consumption in other markets; these are known as 'indirect' effects. For example, a policy which improves insulation in buildings will increase activity in the construction sector and reduce the consumption of oil and gas for space heating, possibly increasing national competitiveness and affecting output in other sectors. These effects are estimated using general equilibrium economic models, which explain the inter-relationships of supply and demand between sectors of an economy. Further discussion of this topic is beyond the scope of this book.

Example of typical costs and benefits for renewable and energy efficient projects are given in Table 6.11.

Example 6.10   Identifying Social Costs and Benefits of a Wind Project or Policy

As part of a feasibility analysis of policies favouring offshore wind energy, a government agency is required to identify all associated costs and benefits. Before doing so, however, the base case must first be established by answering the question: what technology would be preferred if wind energy did not proceed? A review of the industry indicates that the best available technology is combined cycle gas turbine (CCGT) generation, which becomes the technology against which wind is assessed. Expert analysis and stakeholder meetings are held to identify costs and benefits, the results of which are summarised in Table 6.12. Importantly, it is assumed there will be no price benefits to consumers because the level of wind penetration will not be sufficient to change gas-fired generation as the market price setter.

Table 6.12   Results of expert analysis and stakeholder consultations on the costs and benefits of offshore wind power compared to CCGT.

Costs	Benefits
All private investment costs including capital, labour and expenditure on related projects such as port upgrades	The marginal value of all energy produced by the turbines
All private maintenance and operational costs	Additional security of supply due to displacement of imported gas
Impacts on commercial and recreational fishing	Climate change mitigation
Any technology subsidies required	Additional employment and/or salaries over base case
Decreases in property values and tourism	Increased regional commercial activity due to port upgrades

6.5.3   Value costs and benefits

This step involves putting a monetary value on all of the costs and benefits identified. This is straightforward where market prices exist. For example, the social value of energy is what individuals are willing to pay for it on the free market. Assuming the policy or project is sufficiently small not to significantly alter supply and demand in the market, the existing market price can be used. Where supply or demand is affected, the prices and quantities must be adjusted taking into account the price elasticities of demand and supply, which measure the change in demand and supply of goods for a unit change in their price. When the price of a good such as energy increases, then less of it will be demanded, with the amount less being determined by its price elasticity of demand. Carbon taxation policies have the effect of increasing energy prices to the consumer, thus decreasing demand. However, because energy is essential for our everyday activities, it is relatively inelastic, meaning that quantities will not vary as greatly with price fluctuations in other, less essential goods; some elasticities of demand for US commercial and residential energy sources are given in Table 6.13. Similarly, when goods become cheaper to produce, suppliers are willing to produce more. Policies such as technology subsidies have the effect of making energy cheaper to produce, thus increasing demand.

Table 6.13   National price elasticities of demand for gas and electricity in the US residential and commercial markets.

	Residential electricity	Commercial electricity	Residential natural gas
Short-run elasticity	−0.24	−0.21	−0.12
Long-run elasticity	−0.32	−0.97	−0.36

Short-term residential gas consumption is more insensitive to price changes than long-term commercial electricity. In the former case, and 8.3% (1 ÷ 0.12) increase in gas prices will result in a 1% decrease in demand.
Source: Bernstein and Griffin (2005).

However, markets may fail; for example, the cost of energy-related air pollution is often not borne by the polluter and its real value is rarely apparent. Similarly, the value of the visual amenity lost because of a new wind farm has no directly observable market value. In this situation proxy or 'shadow' prices must be established in an attempt to value the impact to society. There are a variety of approaches adopted, some of which are outlined as follows:

- *Revealed preference techniques* are techniques that use the data on consumer purchasing behaviours to reveal their preference (and the price they would be willing to pay) for similar non-traded goods. A number of relevant techniques exist. Hedonic pricing models break the value of a good into its constituent parts; for example, the impact of wind farms on property prices can be statistically estimated using house price and wind farm data. Averted expenditure techniques identify costs which are avoided as a result of the project or policy. It employs data on purchases of goods which are used to manage impacts in the absence of the policy. For example, societal expenditure on air filters and respiratory devices could be used to value policies which reduce particulate emissions.
- *Stated preference techniques* involve surveying stakeholders to establish how much they would agree to pay for avoiding an adverse impact or, alternatively, how much they would ask as compensation to accept the impact. The former is referred to as willingness to pay (WTP) and the latter willingness to accept (WTA). Stakeholders can also be asked to make trade-offs between different alternatives from which their preferences can be established; this is known as 'contingent valuation'. The methods include direct surveys, bidding games and expert interviews.

Shadow price factors (SPF) can be used to convert financial (private) costs to social ones. For example, where a project requires labour and will result in a reduction in unemployment, then the opportunity cost to society of labour is not the full cost of their wages, because some workers would not have been working in the alternative scenario. Here, the cost of labour to the project can be multiplied by a shadow price factor (of less than 1) to convert the private cost to the lower cost to society. SPFs vary both spatially and temporally depending on national and regional economic conditions pertaining at the time of assessment.

Example 6.11   Valuing Costs and Benefits

Approaches to valuing the costs and benefits identified in Example 6.10 are outlined in Table 6.14.

Table 6.14   Approaches to valuing costs and benefits for offshore wind turbines.

Costs	Valuation approach	Benefits	Valuation approach
All private investment costs including capital, labour and expenditure on related projects such as port upgrades	Private financial costs are obtained from the industry from similar projects or by tender	The marginal value of all energy produced by the turbines	Obtain the market value for electricity and estimate quantity of electricity produced
All private maintenance and operational costs	Same as above	Additional security of supply due to displacement of imported gas	Willingness-to-pay surveys and macroeconomic models simulating supply interruptions
Impacts on commercial and recreational fishing	Literature review, impacts of similar developments such as offshore platforms or man-made reefs on fisheries	Climate change mitigation	Use environmental life cycle assessment to estimate emissions savings and apply social cost of carbon or market value from appropriate emissions trading scheme (be careful that cost of carbon is not included in market value of electricity above)
Any technology subsidies required	Estimated using financial analysis to identify capital subsidies, fixed FITs, premium FITs	Additional employment and/or salaries over base case	Apply shadow price factor to private labour costs
Decreases in property values and tourism	Literature review, attitudinal surveys, hedonic pricing	Increased regional commercial activity due to port upgrades	Literature review of impacts of similar upgrades elsewhere; survey of local industry

6.5.4   Discount the costs and benefits

Once values have been applied to costs and benefits, they are discounted in the normal way. The only difference with the method outlined in Chapter 4 is that an appropriate social discount factor, as opposed to a private discount factor should be applied. There are a number of ways in which social discount rates are chosen including the after-tax return on government bonds as well as measures of the social opportunity cost of capital and social rate of time preference. Many national departments of finance (treasuries) publish discount rates which are recommended for discounting public projects. In Ireland, for example, the Irish National Development Finance Agency (NDFA) recommends 5.9% (nominal) for 10-year projects.

Table 6.15 Declining discount rates (real) recommended in the HM Treasury Green Book for use on public programmes and projects.

	Period (yr)					
	0–30	31–75	76–125	126–200	201–300	300+
Discount rate (%)	3.5	3.0	2.5	2.0	1.5	1.0

We observed in Chapter 4 that distant discounted cash flows have almost no value. This represents a problem in social CBA in that the welfare of future generations is largely ignored. For example, a nuclear facility will have significant decontamination costs at the end of its lifespan (40 years or more), whereas a hydroelectric dam can last for hundreds of years. In the former case the costs to future generations will be underestimated; in the latter benefits will be. One way of mitigating this problem is to apply declining discount rates to different time periods so that the policy maker will focus more on improving social welfare in the distant future. For example, the UK Treasury's Green Book recommends the application of declining discount rates to projects which have long-term impacts of over 30 years; these are shown in Table 6.15.

Example 6.12 Declining Discount Rate

A hydroelectric dam has a projected economic lifespan of 100 years. If the total initial costs of the project are €550m and the net annual benefits are €20m, determine the project's NPV using the declining discount rates recommended by the HM Treasury Green Book. Assume an overnight build.

Expanding Equation 4.22 for estimating an NPV using declining discount rates we get

$$NPV = \sum_{n=0}^{N} \frac{F_n}{\prod_{i=1}^{N} (1 + d_i)^n}$$

$$= F_0 + \sum_{n=1}^{30} \frac{F_n}{(1 + d_1)^n} + \sum_{n=31}^{75} \frac{F_n}{(1 + d_2)^n} + \sum_{n=76}^{100} \frac{F_n}{(1 + d_3)^n}$$

where d_1, d_2 and d_3 are the declining discount rates in Table 6.15 for the periods $0 - 30$, $31 - 75$ and $76 - 125$ years respectively. Inserting the capital costs (F_0) of €550m and net annual benefits (F_n) of €20m we get

$$NPV = -550 + \sum_{n=1}^{30} \frac{20}{(1 + 0.035)^n} + \sum_{n=31}^{75} \frac{20}{(1 + 0.030)^n} + \sum_{n=76}^{100} \frac{20}{(1 + 0.025)^n}$$

$$= -550 + 368 + 202 + 58 = €78m$$

6.5.5   Interpret results

Once the values of the discounted costs and benefits have been estimated over the lifespan of the project or policy a number of parameters can be calculated in the same way as for private projects as explained in Chapter 4. Two common metrics used to represent the attractiveness of a project are the NPV and the CBR. The NPV is calculated using

$$\text{NPV} = \sum_{n=0}^{N} \frac{(B_n - C_n)}{(1 + d)^n} \tag{6.10}$$

where C_n is the value of incremental project costs year n, B_n is the value of project benefits year n, N is the analysis period, and d is the annual discount rate.

The CBR is the ratio of the incremental discounted benefits to costs and where positive, it indicates a desirable project. CBR is given by

$$\text{CBR} = \frac{\text{PV}(B_n)}{\text{PV}(C_n)} = \frac{\left[\sum_{n=0}^{N} B_n / (1 + d)^n \right]}{\left[\sum_{n=0}^{N} C_n / (1 + d)^n \right]} \tag{6.11}$$

where PV denotes present value, B_n is the value of incremental project benefits year n, C_n is the project cost year n, N is the analysis period and d is the annual discount rate.

6.5.6   Assess who bears the costs and benefits

For any large energy project, there will be winners and losers. Although such projects may have a positive CBR overall, for some sub-groups and individuals it may be negative. For example, the construction of a hydroelectric dam may result in cheaper and more secure electricity for society as a whole, but communities in flooded valleys, fishermen and those in the tourist industry may be worse off unless compensated in some way. However, it is often possible to redistribute the benefits, so that losers are compensated (e.g. through taxes or transfer payments) and at least one person is better off without making anyone worse off: in this situation the project represents a potential Pareto improvement. In practice, however, projects with positive CBRs often proceed even though it is not always possible to compensate all losers.

Distributional effects are diverse and the groups affected must be classified in different ways depending on the policy in question. Groups may be characterised spatially, for example, where energy projects have environmental impacts in their immediate vicinities – wind farms create flicker and noise for adjacent communities. Different socioeconomic groups may be affected; for example, subsidies for domestic microgeneration systems may favour larger houses with high incomes although the subsidy is spread across all energy consumers, thus representing a transfer to the more affluent. Economic sectors may be affected differently; offshore wind may affect the fishing industry. Furthermore, welfare gains and losses can be weighted differently for different groups depending on policy and societal preferences. For example, benefits to consumers may be given more weighting than those to producers. CBA must attempt to identify and quantify these distributional effects and, where deemed appropriate, weight them according to societal preferences.

6.5.7 Uncertainty

Social CBA, such as financial project appraisal (dealt with in Chapter 4), is a forecasting exercise employing a variety of assumptions and data from different sources. There is, therefore, significant uncertainty inherent in any such analysis undertaken and it is, therefore, important to understand the extent of this uncertainty through risk assessment. The methods described in Section 4.3.8 can be applied to assess the probability of the project or policy performing as planned. These techniques include sensitivity analysis, scenario analysis and Monte Carlo simulation.

6.5.8 Make decision

Finally, the policy or project is deemed desirable it if meets the NPV or CBR thresholds (≥ 0 and ≥ 1 respectively), it meets the Kaldor–Hicks criterion (beneficiaries can compensate losers) and risks are acceptable.

6.6 Case studies

6.6.1 Marginal abatement costs of emission mitigation options in a building estate

A national public sector estate management agency has been given a target of reducing total GHG emissions from all public building by 20% over 6 years (2014–2020). In order to identify the most cost-effective emissions mitigation options it commissions a study which estimates the marginal abatement costs of all feasible carbon-reduction options; once identified each will be subject to more detailed analysis. Marginal abatement cost analysis requires estimations of incremental cost and emissions due to investment, maintenance, operation and decommissioning (i.e. over-and-above the 'business-as-usual' case). The following method is adopted:

1. categorise estate buildings by their level of mechanical and electrical services because this largely determines energy use and, therefore, emissions;

2. develop a CO_2-eq baseline for the estate from energy use and emission intensity data for the most recent year;
3. identify the set of available abatement methods and technologies and their characteristics in terms of fuel inputs, efficiencies, costs, emissions and deployment potential in each building;
4. estimate the incremental costs and emissions when these technologies are deployed to their full feasible potential; and
5. identify the technologies with the lowest MACs for further detailed analysis.

CO_2 only is considered while manufacturing and decommissioning costs are ignored because these are small in the overall LCC for an initial study.

A desktop exercise is first undertaken using existing records, which quantifies fuel consumption by property, gives the gross floor areas (GFA) and provides information on the level of mechanical and electrical servicing in each building. Office level of servicing is defined as

- *Naturally ventilated open plan office*, which has no forced ventilation or air conditioning and typically uses perimeter heating;
- *Naturally ventilated cellular office*, which is similar to the aforementioned but has cellular offices rather than an open-plan layout; and
- *Air-conditioned* in which the majority of working areas are climate controlled using a heating, ventilation and air conditioning (HVAC) system.

Audits of a representative sample of properties were undertaken to gather the data necessary for assessing base case performances as well as estimating the performances of the energy efficiency measures selected. Data were collected on

- the penetration of existing oil and gas boilers as well as their ages and efficiencies;
- the percentages and categories of properties suitable for each of the energy-efficient measures identified;
- the performance of existing mechanical and electrical systems, so that energy-efficient savings can be estimated.

The results of these studies are used to establish energy end-use emissions factors and total estate emissions. Unit fuel emission factors and costs are corrected by boiler efficiencies to give heat emission factors and costs as shown in Table 6.16. The national emissions factor for electricity end-use is 0.54 kg CO_2/kWh.

Unit energy costs and emissions factors from Table 6.16 are combined with estate gross floor area data to estimate costs and emissions for each category of office as shown in Table 6.17. It can be seen under 'energy end use' that heating uses more energy than electricity for naturally ventilated offices, which

Table 6.16   Heat emission factors and costs for the different boiler systems used in the estate.

Boiler system	Proportion of estate (%)	Efficiency (%)	Fuel unit emissions factor (kg/CO_2/kWh)	Heat emissions factor (kg/CO_2/kWh)	Fuel unit cost (c/kWh)	Heat unit cost (c/kWh)
Existing natural gas	50	78	0.20	0.26	3.8	4.9
Existing heating oil	50	70	0.28	0.40	9.9	14.1
Weighted average		*74*	*0.24*	*0.33*	*6.9*	*9.5*

Table 6.17   Floor areas, electrical and heating energy end use, operational emissions and operating costs for each office type in the total estate.

	(m^2)	Electricity – per unit area (kWh/m^2/annum)	Heat – per unit area (kWh/m^2/annum)	Total (MWh/annum)	Per unit area (kgCO_2/m^2/annum)	Total (tCO_2/annum)	Per unit area (€/m^2/annum)	Total (€/annum)
Naturally ventilated open plan offices	363,300	81	166	89,735	98	35,709	24.71	8,977,952
Naturally ventilated cellular offices	337,350	83	138	74,554	90	30,418	22.27	7,511,702
Air-conditioned	181,650	227	175	73,023	180	32,713	41.63	7,561,903
Total	*882,300*			*237,313*		*98,840*		*24,051,557*

make up the majority of the estate (~80%), but that electricity consumption dominates for air-conditioned offices. The higher energy intensity of this later office type means that emissions and costs are equally spread across all three office categories.

Government services are not projected to grow due, in the medium term, to a poor national macroeconomic backdrop. No major technological trends which will significantly alter energy use or emissions are foreseen under normal circumstances. For these reasons energy consumption and costs are estimated to remain unchanged in the BAU scenario.

A meeting of estate facility managers and energy consultants is convened to establish what energy efficiency and/or carbon mitigation measures should be considered in the study. The following measures were shortlisted.

- *Replace Existing Oil Boilers with New Condensing Gas Boilers.* Half of the estate is heated using oil boilers a proportion of which could be replaced by new condensing gas boilers with operational efficiencies of 92%; this will require new connections to the gas grid.
- *Upgrade Old Oil and Gas Boiler to Condensing Boilers.* The vast majority of existing oil and gas boilers are of the old, non-condensing type, and these could be upgraded to condensing technology with operational efficiencies of 92%.

- *Heating Control Maintenance and Upgrade.* A wide variety of heating control systems is used with differing levels of sophistication. Many could be upgraded by improving zoning and heating system controls.
- *HVAC Maintenance and Upgrade.* Existing HVAC systems are old and can be upgraded and better maintained to make them more energy efficient.
- *PV Generation in All Air-Conditioned Offices.* Many of the air-conditioned office types are large and have space for large, accessible PV arrays on their flat roofs.
- *Improve Lighting Efficiency.* Although existing lighting systems are representative of industry best practice, they could be upgraded to state-of-the-art LED technology.

The percentage of the estate suitable for each of these measures, their capital costs and energy savings are estimated by the energy consultants in collaboration with the facility managers based on the audit findings (Table 6.18).

An assessment of other costs such as insurances, maintenance and administration was undertaken but were found not to represent significant for the energy efficiency measures considered; these were, therefore, omitted from the incremental cost analysis. Incremental emissions and costs were then estimated and discounted for each year of operation over their expected lifespan for each mitigation measure. These results, along with the marginal abatement costs (calculated using Equation 6.5) are shown in Table 6.19.

Figure 6.6 shows the marginal abatement cost curve for the abatement measures analysed. Deployment potential on the horizontal axis should be treated with caution because the avoided emissions for each measure are calculated separately; the combined effects of multiple technologies are likely to give lower cumulative emissions. From this figure, it can be seen that three

Table 6.18 Emission mitigation measures identified for the estate along with their deployment potentials (as a percentage of gross estate floor area) and capital costs.

Mitigation measure	Proportion of estate suitable	Energy saving	Capital cost (€)
Replace oil boilers with new condensing gas boilers	8%	Increased boiler efficiency to 92%	9,175,920
HVAC maintenance and upgrade	21%	17% reduction in heat consumption	3,271,568
Heating control maintenance and upgrade	40%	8% reduction in heat consumption	3,882,120
Upgrade old oil and gas boiler to condensing boilers	80%	Increased boiler efficiency to 92%	69,172,320
PV generation in all air-conditioned offices	30%	8029 MWh electrical output per annum	19,073,250
Improve lighting efficiency	100%	5% reduction in electrical consumption	9,705,300

Table 6.19 Marginal abatement costs for each mitigation measure based on discounted incremental emissions and cost savings over the investment lifespan.

Mitigation measure	Operational emissions		Cum op emissions savings (tCO$_2$/annum)	Operational costs		Capital costs (€)	Lifespan (years)	Incremental life cycle costs (€)	Discounted lifecycle emissions savings (tCO$_2$)	Marginal abatement costs (€/tCO$_2$)
	Total (tCO$_2$/annum)	Savings (tCO$_2$/annum)		Total (€/annum)	Savings (€/annum)					
None (business-as-usual)	98,840.00			24,051,557.31						
Replace oil boilers with new condensing gas boilers	96,805.00	2,035.00	2,035.00	22,946,174.18	1,105,383	9,175,920	15	(1,559,836)	19,764	(79)
HVAC maintenance and upgrade	97,063.83	1,776.17	3,811.17	23,537,119.86	514,437	3,271,568	10	(514,736)	13,073	(39)
Heating control maintenance and upgrade	97,381.68	1,458.32	5,269.50	23,629,197.37	422,360	3,882,120	15	(219,945)	14,164	(16)
Upgrade old oil and gas boiler to condensing boilers	86,503.00	12,337.00	17,606.50	18,110,234.80	5,941,323	69,172,320	15	11,468,716	119,820	96
PV generation in all air-conditioned offices	94,504.05	4,335.95	21,942.45	23,168,375.01	883,182	19,073,250	25	7,783,216	55,428	140
Improve lighting efficiency	96,175.80	2,664.20	24,606.65	23,508,916.86	542,640	9,705,300	10	5,711,419	19,609	291

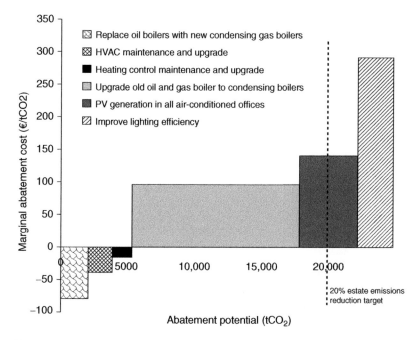

Figure 6.6   Marginal abatement cost curve for the mitigation measures for the estate.

measures have negative MACs and, therefore, result in LCC savings *and* emissions reductions for the estate. The measure with the greatest mitigation impacts is the upgrade of all existing boilers with a MAC of 96 €/tCO$_2$. Lighting improvements has the highest MAC, most likely because existing lighting systems are already efficient so that the incremental emissions and energy savings are small compared to the investment involved. The total deployment potential of the individual measures (ignoring the fact that combined cumulative savings will be lower) results in avoided emissions of almost 25,000 tCO$_2$-eq/annum or 25% of annual emissions.

The estate agency shortlists all technologies except lighting for further detailed analysis. This will involve

- more detailed LCC analysis;
- a full environmental LCA based on CO$_2$-equivalent incorporating embodied, maintenance and decommissioning emissions; and
- the effects of technology combinations on costs and emissions.

6.6.2   PV feed-in-tariff design

National policy makers wish to achieve a 30% penetration of PV systems in the residential sector and have commissioned a government energy agency to design a suitable FIT and assess the total cost of such a policy.

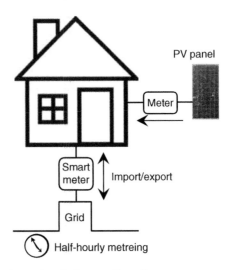

Figure 6.7   PV and smart metering configuration.

The most suitable FIT type is dependent on the metering and connection configuration that will be employed and the metered data available. The assessment is being undertaken during the implementation of a national residential smart-metering programme, which will make half-hourly household import and export data available, thus giving significant flexibility in the type of FIT design to adopt. The standard dwelling grid interface comprises a grid connection monitored with a smart meter and a metered PV array connected onto the domestic electrical distribution system (see Figure 6.7). For the purpose of transparency, it is decided to apply a fixed FIT to all electricity which is produced by the PV system. The incremental system revenues thus depend on:

- electricity imports displaced by the system at the relevant domestic electricity tariff;
- electricity exported which is compensated at the grid marginal cost of production (the 'best new entrant' price); and
- electricity produced by the PV system which is compensated at the FIT (see Figure 6.8).

Incremental costs are the capital cost of the system and minor maintenance costs.

The approach adopted by the agency in the design of an economically efficient FIT involves

1. developing a half-hourly demand financial model;
2. using this model to simulate the performances of PV systems in of dwellings, which are representative of the national housing stock;
3. determining the distribution of export FITs required to give NPVs of zero; and

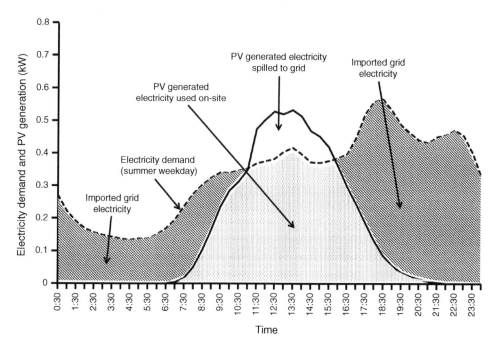

Figure 6.8   Sample daily profile showing household demand, PV supply and electricity import and export.

4. choosing a suitable tariff which will incentivise 30% of the population to invest.

Data sources include:

- a year of half-hourly electricity domestic demand profiles form a representative sample of 500 smart metres;
- a year of validated half-hourly electricity supply from a 1.5 kWp domestic PV system representative of the best available technology (a previous study by the agency has indicated that the optimum size of PV system for the national housing stock is 1.5 kWp);
- industry data on PV capital and maintenance costs together with published technology learning rates;
- average domestic energy tariffs from the national statistics provider; and
- real personal discount rates for similar investments reported in the international literature.

For each dwelling, the FIT required to make the investment financially attractive to the investor is obtained by setting the NPV of the PV system to zero and solving for the relevant FIT. A financial model is developed according to Equation 4.19, which relates NPV, capital cost and energy-related revenues:

$$\text{NPV} = R_M - C_M \tag{6.12}$$

where NPV is the net present value (€), R_M is the present value of total revenue (€) and C_M is the present value of total cost (€).

Revenue depends primarily on the amount of electricity produced by the PV system (system yield) and is a function of the electricity load profile of the PV owner, import and export tariffs, as well as the FIT. The total revenue in year m is calculated as

$$R_m = \sum_{n=1}^{N} (\alpha E_{A,n} + \beta E_{G,n} + \gamma E_{E,n}) \qquad (6.13)$$

where R_m is the total revenue in year m (€), α is the household domestic electricity tariff (€/kWh), $E_{A,n}$ is avoided electricity imports (kWh) in half-hour period n, β is the FIT giving an NPV of zero (€/kWh), $E_{G,n}$ is total electricity generated (kWh), γ is the export electricity tariff, $E_{E,n}$ is electricity exported (kWh) in half-hour period n and N is the number of half hours in 1 year.

The present value of total revenue is calculated as

$$R_M = \sum_{m=1}^{M} \left[\sum_{n=1}^{N} \left(\alpha E_{A,n} + \beta E_{G,n} + \gamma E_{E,n} \right) \right] (1+d)^m \qquad (6.14)$$

where R_m is the present value of total revenue over the investment lifespan (€), M is the investment lifespan (years) and d is the household discount rate.

The present value of total cost is calculated as

$$C_M = C_0 + C_{O\&M,m} \sum_{m=1}^{M} (1+d)^{-m} \qquad (6.15)$$

where C_0 is the capital cost of the system in year zero and $C_{O\&M,m}$ is the operation and maintenance cost in year m.

Substituting R_M (Equation 6.14) and C_M (Equation 6.15) into Equation 6.12, setting NPV to zero and rearranging to solve for β gives the FIT required to incentivise investment as shown in Equation 6.16.

$$\beta = \frac{1}{\sum\limits_{m=1}^{M} \left[\sum\limits_{n=1}^{n=N} E_{G,n} \right] (1+d)^{-n}}$$
$$\times \left[C_0 + \sum_{m=1}^{M} \left(C_{O\&M,m} - \left[\sum_{n=1}^{N} \left(\alpha E_{A,n} + \gamma E_{E,n} \right) \right] \right) (1+d)^{-m} \right] \qquad (6.16)$$

A computer model is developed based on Equation 6.16 and using the parameters in Table 6.20. FITs are then established for each of the 500 dwellings by simulating energy imports and exports using half-hourly household electricity demand and PV supply profiles. A frequency distribution of FITs required to achieve an NPV of zero for each dwelling is shown in

Table 6.20   Key parameter values and sources.

Capital cost	C_0	3250 €/PV system	Industry survey
Maintenance cost	$C_{O\&M}$	35 €/annum	Industry survey
Lifespan	M	25 years	Literature
Domestic electricity tariff	α	0.2 €/kWh	National statistics
Export electricity tariff	γ	0.05 €/kWh	
Personal discount rate	d	20%	Literature

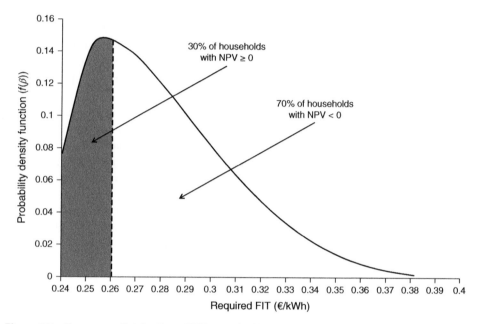

Figure 6.9   Frequency distribution of FITs required by dwellings to give an NPV of zero, thus incentivising the uptake of the 1.5 kWp PV system.

Figure 6.9. This shows that A FIT of 0.261 €/kWh is required to incentivise 30% of households to invest in the technology.

A more complete analysis would involve

- optimising system sizes for individual dwellings to minimise FIT, increasing economic efficiency;
- incorporating parameters that approximate inefficiencies such as overshadowing, sub-optimal panel orientation and panel efficiency losses over time;
- consideration of possible changes in electricity demand profiles over the 25-year lifespan;
- making necessary adjustments for learning rates to coincide with the date of commencement of the FIT scheme; and
- analysis to deal with uncertainties in key parameters including personal discount rates and electricity tariffs.

6.7 Conclusions

Governments around the world are moving to decarbonise their energy systems through a combination of regulations, capital subsidies, FITs and carbon pricing. These policy instruments must be properly designed to incentivise private investors, offer value-for-money to society while ensuring all benefits are evenly distributed. The large number of abatement technology options available to the policy maker can make it difficult to know where to begin. MCDA techniques can be used to shortlist available options based on financial and non-financial criteria using approximate data for further detailed analysis. Emissions policy targets can be chosen using marginal abatement costs and deployment potentials for different technologies; however, care must be taken in interpreting MAC cost curves, so that multiple technology interactions are considered in policy formulation. Achieving technology targets requires the design of efficient financial incentives such as capital subsidies and FITs using NPV and LCOE approaches, respectively. The methods adopted should consider sectoral and macroeconomic effects, although these are not described here. Policy options can be tested for their efficiency and equity using CBA. There is no one approach to the design of policy instruments and the process is not a sequential process. An iterative approach is typically involved using available tools and data while satisfying political and time constraints. Economic performance may not be the only or even the most important consideration.

References

Bernstein, M. and Griffin, J. (2005) Regional Differences in the Price-Elasticity of Demand For Energy. Prepared for the National Renewables Energy Laboratory by RAND Infrastructure, Safety and Environment (ISE). RAND Corporation.

BPIE (2011) *Europe's buildings under the microscope: A country-by-country review of the energy performance of buildings.* Brussels, Buildings Performance Institute Europe (BPIE).

Bullard, C., Penner, P., & Pilati, D. (1978) Net energy analysis – Handbook for combining process and input-output analysis. *Resources and Energy*, **1**, 267–313.

Crawford, R. (2008) Validation of a hybrid life-cycle inventory analysis method. *Journal of Environmental Management*, **88**, 496–506.

Hammond, G. and Jones, C. (2011) Inventory of Carbon and Energy (ICE) Database, Version 2.0. University of Bath.

Hood, C. (2011) *Summing Up the Parts – Combining Policy Instruments with Least-cost Climate Mitigation Strategies.* International Energy Agency, Paris.

IEA (2013) *World Energy Outlook 2011: IEA analysis of fossil fuel subsidies.* International Energy Agency, Paris.

IEA (2014) Policies & Measures Databases. Available at www.iea.org/policiesand measures/. Accessed 21 Oct 2014.

IPCC, (2007: Climate change 2007: Synthesis report. *Contribution of Working Groups I, II and III to the Fourth Assessment Report of the Intergovernmental Panel on Climate Change* [Core Writing Team, Pachauri, R.K and Reisinger, A. (eds.)]. IPCC, Geneva, Switzerland, 104 pp.

IPCC (2013) Climate change 2013: The physical science basis. In: *Working Group I Contribution to the Fifth Assessment Report of the Intergovernmental Panel on Climate Change* (eds T.F. Stocker, D. Qin, G.-K. Plattner, M.M.B. Tignor, S.K. Allen, J. Boschung, A. Nauels, Y. Xia, V. Bex, & P.M. Midgley). Cambridge University Press.

McKinsey and Company (2009) *Pathways to a low-carbon economy: Version 2 of the global greenhouse gas abatement cost curve.* Company, McKinsey and.

S Suh, G Huppes, Methods for life cycle inventory of a product, *Journal of Cleaner Production*, Volume **13**, Issue 7, 2005, Pages 687-697

Appendix A

Table of Discount Factors

Period	Discount Rate %																			
	1		2		3		4		5		6		7		8		9		10	
1	0.990	0.990	0.980	0.980	0.971	0.971	0.962	0.962	0.952	0.952	0.943	0.943	0.935	0.935	0.926	0.926	0.917	0.917	0.909	0.909
2	0.980	1.970	0.961	1.942	0.943	1.913	0.925	1.886	0.907	1.859	0.890	1.833	0.873	1.808	0.857	1.783	0.842	1.759	0.826	1.736
3	0.971	2.941	0.942	2.884	0.915	2.829	0.889	2.775	0.864	2.723	0.840	2.673	0.816	2.624	0.794	2.577	0.772	2.531	0.751	2.487
4	0.961	3.902	0.924	3.808	0.888	3.717	0.855	3.630	0.823	3.546	0.792	3.465	0.763	3.387	0.735	3.312	0.708	3.240	0.683	3.170
5	0.951	4.853	0.906	4.713	0.863	4.580	0.822	4.452	0.784	4.329	0.747	4.212	0.713	4.100	0.681	3.993	0.650	3.890	0.621	3.791
6	0.942	5.795	0.888	5.601	0.837	5.417	0.790	5.242	0.746	5.076	0.705	4.917	0.666	4.767	0.630	4.623	0.596	4.486	0.564	4.355
7	0.933	6.728	0.871	6.472	0.813	6.230	0.760	6.002	0.711	5.786	0.665	5.582	0.623	5.389	0.583	5.206	0.547	5.033	0.513	4.868
8	0.923	7.652	0.853	7.325	0.789	7.020	0.731	6.733	0.677	6.463	0.627	6.210	0.582	5.971	0.540	5.747	0.502	5.535	0.467	5.335
9	0.914	8.566	0.837	8.162	0.766	7.786	0.703	7.435	0.645	7.108	0.592	6.802	0.544	6.515	0.500	6.247	0.460	5.995	0.424	5.759
10	0.905	9.471	0.820	8.983	0.744	8.530	0.676	8.111	0.614	7.722	0.558	7.360	0.508	7.024	0.463	6.710	0.422	6.418	0.386	6.145
11	0.896	10.368	0.804	9.787	0.722	9.253	0.650	8.760	0.585	8.306	0.527	7.887	0.475	7.499	0.429	7.139	0.388	6.805	0.350	6.495
12	0.887	11.255	0.788	10.575	0.701	9.954	0.625	9.385	0.557	8.863	0.497	8.384	0.444	7.943	0.397	7.536	0.356	7.161	0.319	6.814
13	0.879	12.134	0.773	11.348	0.681	10.635	0.601	9.986	0.530	9.394	0.469	8.853	0.415	8.358	0.368	7.904	0.326	7.487	0.290	7.103
14	0.870	13.004	0.758	12.106	0.661	11.296	0.577	10.563	0.505	9.899	0.442	9.295	0.388	8.745	0.340	8.244	0.299	7.786	0.263	7.367
15	0.861	13.865	0.743	12.849	0.642	11.938	0.555	11.118	0.481	10.380	0.417	9.712	0.362	9.108	0.315	8.559	0.275	8.061	0.239	7.606
16	0.853	14.718	0.728	13.578	0.623	12.561	0.534	11.652	0.458	10.838	0.394	10.106	0.339	9.447	0.292	8.851	0.252	8.313	0.218	7.824
17	0.844	15.562	0.714	14.292	0.605	13.166	0.513	12.166	0.436	11.274	0.371	10.477	0.317	9.763	0.270	9.122	0.231	8.544	0.198	8.022
18	0.836	16.398	0.700	14.992	0.587	13.754	0.494	12.659	0.416	11.690	0.350	10.828	0.296	10.059	0.250	9.372	0.212	8.756	0.180	8.201
19	0.828	17.226	0.686	15.678	0.570	14.324	0.475	13.134	0.396	12.085	0.331	11.158	0.277	10.336	0.232	9.604	0.194	8.950	0.164	8.365
20	0.820	18.046	0.673	16.351	0.554	14.877	0.456	13.590	0.377	12.462	0.312	11.470	0.258	10.594	0.215	9.818	0.178	9.129	0.149	8.514
21	0.811	18.857	0.660	17.011	0.538	15.415	0.439	14.029	0.359	12.821	0.294	11.764	0.242	10.836	0.199	10.017	0.164	9.292	0.135	8.649
22	0.803	19.660	0.647	17.658	0.522	15.937	0.422	14.451	0.342	13.163	0.278	12.042	0.226	11.061	0.184	10.201	0.150	9.442	0.123	8.772
23	0.795	20.456	0.634	18.292	0.507	16.444	0.406	14.857	0.326	13.489	0.262	12.303	0.211	11.272	0.170	10.371	0.138	9.580	0.112	8.883
24	0.788	21.243	0.622	18.914	0.492	16.936	0.390	15.247	0.310	13.799	0.247	12.550	0.197	11.469	0.158	10.529	0.126	9.707	0.102	8.985
25	0.780	22.023	0.610	19.523	0.478	17.413	0.375	15.622	0.295	14.094	0.233	12.783	0.184	11.654	0.146	10.675	0.116	9.823	0.092	9.077
26	0.772	22.795	0.598	20.121	0.464	17.877	0.361	15.983	0.281	14.375	0.220	13.003	0.172	11.826	0.135	10.810	0.106	9.929	0.084	9.161
27	0.764	23.560	0.586	20.707	0.450	18.327	0.347	16.330	0.268	14.643	0.207	13.211	0.161	11.987	0.125	10.935	0.098	10.027	0.076	9.237
28	0.757	24.316	0.574	21.281	0.437	18.764	0.333	16.663	0.255	14.898	0.196	13.406	0.150	12.137	0.116	11.051	0.090	10.116	0.069	9.307
29	0.749	25.066	0.563	21.844	0.424	19.188	0.321	16.984	0.243	15.141	0.185	13.591	0.141	12.278	0.107	11.158	0.082	10.198	0.063	9.370
30	0.742	25.808	0.552	22.396	0.412	19.600	0.308	17.292	0.231	15.372	0.174	13.765	0.131	12.409	0.099	11.258	0.075	10.274	0.057	9.427

continued overleaf…

Table of Discount Factors

<table>
<tr><th rowspan="2">Period</th><th colspan="20">Discount Rate %</th></tr>
<tr><th colspan="2">11</th><th colspan="2">12</th><th colspan="2">13</th><th colspan="2">14</th><th colspan="2">15</th><th colspan="2">16</th><th colspan="2">17</th><th colspan="2">18</th><th colspan="2">19</th><th colspan="2">20</th></tr>
<tr><td>1</td><td>0.901</td><td>0.901</td><td>0.893</td><td>0.893</td><td>0.885</td><td>0.885</td><td>0.877</td><td>0.877</td><td>0.870</td><td>0.870</td><td>0.862</td><td>0.862</td><td>0.855</td><td>0.855</td><td>0.847</td><td>0.847</td><td>0.840</td><td>0.840</td><td>0.833</td><td>0.833</td></tr>
<tr><td>2</td><td>0.812</td><td>1.713</td><td>0.797</td><td>1.690</td><td>0.783</td><td>1.668</td><td>0.769</td><td>1.647</td><td>0.756</td><td>1.626</td><td>0.743</td><td>1.605</td><td>0.731</td><td>1.585</td><td>0.718</td><td>1.566</td><td>0.706</td><td>1.547</td><td>0.694</td><td>1.528</td></tr>
<tr><td>3</td><td>0.731</td><td>2.444</td><td>0.712</td><td>2.402</td><td>0.693</td><td>2.361</td><td>0.675</td><td>2.322</td><td>0.658</td><td>2.283</td><td>0.641</td><td>2.246</td><td>0.624</td><td>2.210</td><td>0.609</td><td>2.174</td><td>0.593</td><td>2.140</td><td>0.579</td><td>2.106</td></tr>
<tr><td>4</td><td>0.659</td><td>3.102</td><td>0.636</td><td>3.037</td><td>0.613</td><td>2.974</td><td>0.592</td><td>2.914</td><td>0.572</td><td>2.855</td><td>0.552</td><td>2.798</td><td>0.534</td><td>2.743</td><td>0.516</td><td>2.690</td><td>0.499</td><td>2.639</td><td>0.482</td><td>2.589</td></tr>
<tr><td>5</td><td>0.593</td><td>3.696</td><td>0.567</td><td>3.605</td><td>0.543</td><td>3.517</td><td>0.519</td><td>3.433</td><td>0.497</td><td>3.352</td><td>0.476</td><td>3.274</td><td>0.456</td><td>3.199</td><td>0.437</td><td>3.127</td><td>0.419</td><td>3.058</td><td>0.402</td><td>2.991</td></tr>
<tr><td>6</td><td>0.535</td><td>4.231</td><td>0.507</td><td>4.111</td><td>0.480</td><td>3.998</td><td>0.456</td><td>3.889</td><td>0.432</td><td>3.784</td><td>0.410</td><td>3.685</td><td>0.390</td><td>3.589</td><td>0.370</td><td>3.498</td><td>0.352</td><td>3.410</td><td>0.335</td><td>3.326</td></tr>
<tr><td>7</td><td>0.482</td><td>4.712</td><td>0.452</td><td>4.564</td><td>0.425</td><td>4.423</td><td>0.400</td><td>4.288</td><td>0.376</td><td>4.160</td><td>0.354</td><td>4.039</td><td>0.333</td><td>3.922</td><td>0.314</td><td>3.812</td><td>0.296</td><td>3.706</td><td>0.279</td><td>3.605</td></tr>
<tr><td>8</td><td>0.434</td><td>5.146</td><td>0.404</td><td>4.968</td><td>0.376</td><td>4.799</td><td>0.351</td><td>4.639</td><td>0.327</td><td>4.487</td><td>0.305</td><td>4.344</td><td>0.285</td><td>4.207</td><td>0.266</td><td>4.078</td><td>0.249</td><td>3.954</td><td>0.233</td><td>3.837</td></tr>
<tr><td>9</td><td>0.391</td><td>5.537</td><td>0.361</td><td>5.328</td><td>0.333</td><td>5.132</td><td>0.308</td><td>4.946</td><td>0.284</td><td>4.772</td><td>0.263</td><td>4.607</td><td>0.243</td><td>4.451</td><td>0.225</td><td>4.303</td><td>0.209</td><td>4.163</td><td>0.194</td><td>4.031</td></tr>
<tr><td>10</td><td>0.352</td><td>5.889</td><td>0.322</td><td>5.650</td><td>0.295</td><td>5.426</td><td>0.270</td><td>5.216</td><td>0.247</td><td>5.019</td><td>0.227</td><td>4.833</td><td>0.208</td><td>4.659</td><td>0.191</td><td>4.494</td><td>0.176</td><td>4.339</td><td>0.162</td><td>4.192</td></tr>
<tr><td>11</td><td>0.317</td><td>6.207</td><td>0.287</td><td>5.938</td><td>0.261</td><td>5.687</td><td>0.237</td><td>5.453</td><td>0.215</td><td>5.234</td><td>0.195</td><td>5.029</td><td>0.178</td><td>4.836</td><td>0.162</td><td>4.656</td><td>0.148</td><td>4.486</td><td>0.135</td><td>4.327</td></tr>
<tr><td>12</td><td>0.286</td><td>6.492</td><td>0.257</td><td>6.194</td><td>0.231</td><td>5.918</td><td>0.208</td><td>5.660</td><td>0.187</td><td>5.421</td><td>0.168</td><td>5.197</td><td>0.152</td><td>4.988</td><td>0.137</td><td>4.793</td><td>0.124</td><td>4.611</td><td>0.112</td><td>4.439</td></tr>
<tr><td>13</td><td>0.258</td><td>6.750</td><td>0.229</td><td>6.424</td><td>0.204</td><td>6.122</td><td>0.182</td><td>5.842</td><td>0.163</td><td>5.583</td><td>0.145</td><td>5.342</td><td>0.130</td><td>5.118</td><td>0.116</td><td>4.910</td><td>0.104</td><td>4.715</td><td>0.093</td><td>4.533</td></tr>
<tr><td>14</td><td>0.232</td><td>6.982</td><td>0.205</td><td>6.628</td><td>0.181</td><td>6.302</td><td>0.160</td><td>6.002</td><td>0.141</td><td>5.724</td><td>0.125</td><td>5.468</td><td>0.111</td><td>5.229</td><td>0.099</td><td>5.008</td><td>0.088</td><td>4.802</td><td>0.078</td><td>4.611</td></tr>
<tr><td>15</td><td>0.209</td><td>7.191</td><td>0.183</td><td>6.811</td><td>0.160</td><td>6.462</td><td>0.140</td><td>6.142</td><td>0.123</td><td>5.847</td><td>0.108</td><td>5.575</td><td>0.095</td><td>5.324</td><td>0.084</td><td>5.092</td><td>0.074</td><td>4.876</td><td>0.065</td><td>4.675</td></tr>
<tr><td>16</td><td>0.188</td><td>7.379</td><td>0.163</td><td>6.974</td><td>0.141</td><td>6.604</td><td>0.123</td><td>6.265</td><td>0.107</td><td>5.954</td><td>0.093</td><td>5.668</td><td>0.081</td><td>5.405</td><td>0.071</td><td>5.162</td><td>0.062</td><td>4.938</td><td>0.054</td><td>4.730</td></tr>
<tr><td>17</td><td>0.170</td><td>7.549</td><td>0.146</td><td>7.120</td><td>0.125</td><td>6.729</td><td>0.108</td><td>6.373</td><td>0.093</td><td>6.047</td><td>0.080</td><td>5.749</td><td>0.069</td><td>5.475</td><td>0.060</td><td>5.222</td><td>0.052</td><td>4.990</td><td>0.045</td><td>4.775</td></tr>
<tr><td>18</td><td>0.153</td><td>7.702</td><td>0.130</td><td>7.250</td><td>0.111</td><td>6.840</td><td>0.095</td><td>6.467</td><td>0.081</td><td>6.128</td><td>0.069</td><td>5.818</td><td>0.059</td><td>5.534</td><td>0.051</td><td>5.273</td><td>0.044</td><td>5.033</td><td>0.038</td><td>4.812</td></tr>
<tr><td>19</td><td>0.138</td><td>7.839</td><td>0.116</td><td>7.366</td><td>0.098</td><td>6.938</td><td>0.083</td><td>6.550</td><td>0.070</td><td>6.198</td><td>0.060</td><td>5.877</td><td>0.051</td><td>5.584</td><td>0.043</td><td>5.316</td><td>0.037</td><td>5.070</td><td>0.031</td><td>4.843</td></tr>
<tr><td>20</td><td>0.124</td><td>7.963</td><td>0.104</td><td>7.469</td><td>0.087</td><td>7.025</td><td>0.073</td><td>6.623</td><td>0.061</td><td>6.259</td><td>0.051</td><td>5.929</td><td>0.043</td><td>5.628</td><td>0.037</td><td>5.353</td><td>0.031</td><td>5.101</td><td>0.026</td><td>4.870</td></tr>
<tr><td>21</td><td>0.112</td><td>8.075</td><td>0.093</td><td>7.562</td><td>0.077</td><td>7.102</td><td>0.064</td><td>6.687</td><td>0.053</td><td>6.312</td><td>0.044</td><td>5.973</td><td>0.037</td><td>5.665</td><td>0.031</td><td>5.384</td><td>0.026</td><td>5.127</td><td>0.022</td><td>4.891</td></tr>
<tr><td>22</td><td>0.101</td><td>8.176</td><td>0.083</td><td>7.645</td><td>0.068</td><td>7.170</td><td>0.056</td><td>6.743</td><td>0.046</td><td>6.359</td><td>0.038</td><td>6.011</td><td>0.032</td><td>5.696</td><td>0.026</td><td>5.410</td><td>0.022</td><td>5.149</td><td>0.018</td><td>4.909</td></tr>
<tr><td>23</td><td>0.091</td><td>8.266</td><td>0.074</td><td>7.718</td><td>0.060</td><td>7.230</td><td>0.049</td><td>6.792</td><td>0.040</td><td>6.399</td><td>0.033</td><td>6.044</td><td>0.027</td><td>5.723</td><td>0.022</td><td>5.432</td><td>0.018</td><td>5.167</td><td>0.015</td><td>4.925</td></tr>
<tr><td>24</td><td>0.082</td><td>8.348</td><td>0.066</td><td>7.784</td><td>0.053</td><td>7.283</td><td>0.043</td><td>6.835</td><td>0.035</td><td>6.434</td><td>0.028</td><td>6.073</td><td>0.023</td><td>5.746</td><td>0.019</td><td>5.451</td><td>0.015</td><td>5.182</td><td>0.013</td><td>4.937</td></tr>
<tr><td>25</td><td>0.074</td><td>8.422</td><td>0.059</td><td>7.843</td><td>0.047</td><td>7.330</td><td>0.038</td><td>6.873</td><td>0.030</td><td>6.464</td><td>0.024</td><td>6.097</td><td>0.020</td><td>5.766</td><td>0.016</td><td>5.467</td><td>0.013</td><td>5.195</td><td>0.010</td><td>4.948</td></tr>
<tr><td>26</td><td>0.066</td><td>8.488</td><td>0.053</td><td>7.896</td><td>0.042</td><td>7.372</td><td>0.033</td><td>6.906</td><td>0.026</td><td>6.491</td><td>0.021</td><td>6.118</td><td>0.017</td><td>5.783</td><td>0.014</td><td>5.480</td><td>0.011</td><td>5.206</td><td>0.009</td><td>4.956</td></tr>
<tr><td>27</td><td>0.060</td><td>8.548</td><td>0.047</td><td>7.943</td><td>0.037</td><td>7.409</td><td>0.029</td><td>6.935</td><td>0.023</td><td>6.514</td><td>0.018</td><td>6.136</td><td>0.014</td><td>5.798</td><td>0.011</td><td>5.492</td><td>0.009</td><td>5.215</td><td>0.007</td><td>4.964</td></tr>
<tr><td>28</td><td>0.054</td><td>8.602</td><td>0.042</td><td>7.984</td><td>0.033</td><td>7.441</td><td>0.026</td><td>6.961</td><td>0.020</td><td>6.534</td><td>0.016</td><td>6.152</td><td>0.012</td><td>5.810</td><td>0.010</td><td>5.502</td><td>0.008</td><td>5.223</td><td>0.006</td><td>4.970</td></tr>
<tr><td>29</td><td>0.048</td><td>8.650</td><td>0.037</td><td>8.022</td><td>0.029</td><td>7.470</td><td>0.022</td><td>6.983</td><td>0.017</td><td>6.551</td><td>0.014</td><td>6.166</td><td>0.011</td><td>5.820</td><td>0.008</td><td>5.510</td><td>0.006</td><td>5.229</td><td>0.005</td><td>4.975</td></tr>
<tr><td>30</td><td>0.044</td><td>8.694</td><td>0.033</td><td>8.055</td><td>0.026</td><td>7.496</td><td>0.020</td><td>7.003</td><td>0.015</td><td>6.566</td><td>0.012</td><td>6.177</td><td>0.009</td><td>5.829</td><td>0.007</td><td>5.517</td><td>0.005</td><td>5.235</td><td>0.004</td><td>4.979</td></tr>
</table>

Index

Note: page numbers in *italic* refer to figures and tables.

abatement cost. *see* marginal abatement
    cost
accelerated depreciation 113
aggregated cash values 116. *see also* net
    present value
aggregation (decision analysis) 151–2
air power. *see* compressed air energy
    storage; wind power
aleatoric uncertainty 134
analytic hierarchy process (AHP)
    (decision analysis) 168–81,
    201
appraisal techniques 2–9.
    see also cost-benefit analysis;
    financial analysis; life cycle
    assessment; multi-criteria decision
    analysis
approximation method (analytic
    hierarchy process) 172–3
artificial systems 57–8
assets, energy-efficiency projects 102,
    112–13, 121
assumptions (computer modelling) 78,
    84, 91–2
attitude-oriented methods (decision
    analysis) 158–60, 201
availability factor 13
availability of energy
    (dispatchability) 8, 210

base year (cash flow) 104–6
benefit-cost analysis. *see* cost-benefit
    analysis
benefit-cost ratio (CBR) 121–3, 230,
    237–8

boilers 28–30, *29*
    cash flow examples 103–4, 109–10,
        114, 118–19
    modelling 53–5, 61–2, 64
    policy examples 223, 240–3

CAES. *see* compressed air energy
    storage
calorific value (energy) 12
capacity factor (CF) 13
    combined cycle gas turbine plant
        131
    feed-in-tariffs 226, 228–9
    gas-fired power plant 220
    hydropower 19
capital budgeting. *see* financial analysis
capital costs. *see* investment costs
capital subsidies 120, 207, 225, 228,
    229
carbon capture and storage
    (CCS) 16–17
carbon dioxide emissions. *see*
    greenhouse gas emissions
carbon taxes 7, 204–7, 223, 233
case studies 6, 9
    cash flow 103–4, 141–2,
        145–6
    computer modelling 83–93
    financial analysis 139–48
    multi-criteria decision
        analysis 189–200
    policy *206*, 238–47
cash flows 9, 134, 148–9
    case studies 103–4, 141–2, 145–6
    classification of 103–4

Renewable Energy and Energy Efficiency: Assessment of Projects and Policies, First Edition.
Aidan Duffy, Martin Rogers and Lacour Ayompe.
© 2015 John Wiley & Sons, Ltd. Published 2015 by John Wiley & Sons, Ltd.
Companion Website: www.wiley.com/go/duffy/renewable

cash flows (*cont'd*)
    computer modelling 53, 59, 83–94, 97
    depreciation 112–14
    discount rates 109–12, 236
    financial analysis 97–8, 100–6
    financial measures 116–17, *137–8*
    internal rate of return 127–30
    life cycle cost 131–2
    payback periods 117–20
    present value 106–9, 123–7, *127*
    projects with unequal lifespans 115–16
    real and nominal prices 104–6
    return on investment 120–3
cash inflow 101–2. *see also* revenues
cash outflow 101–2. *see also* operation and maintenance costs
CBA. *see* cost-benefit analysis
CBR (cost-benefit ratio) 121–3, 230, 237–8
CCS (carbon capture and storage) 16–17
CF. *see* capacity factor
CHP. *see* combined heat and power
coal-fired power plants 15–17, 204
combination (combi) boilers 29
combined cycle coal-fired plants 16
combined cycle gas turbine (CCGT) plants *14*, 15, 131–2, 232–3
combined heat and power (CHP) 34–9
    cash flow 102, 103–4
    engines/ turbines 37
    computer modelling 53, 65–6
    micro-CHP 36, *36–7*
combined heat, power and cooling 38–9
combustion 3, 12
community impacts. *see* societal impacts of energy production
company (corporation) tax 113–14
complimentary projects 100
compressed air energy storage (CAES) 42–3, *43*
    computer modelling 90–3, *94*
    net present value 124–5, *125–6*
concordance analysis (decision analysis) 181–9

condensation 12, 38
condensing boilers 29, 240, *241–3*
conjunctive method (decision analysis) 155–6
consumer price index (CPI) 105–6
consumption of energy 13. *see also* use of energy
contingent projects 100
contingent valuation 234
conversion of energy 5, 12–13, 15, 82, 203, *204*
cooling systems 38–9, 87–90
corporation tax 113–4
cost metrics 116–7
cost of energy 2–3, 6–7, 45–6
    boilers 29
    coal-fired power plants 16
    combined heat and power 35, 36, 38
    computer modelling 53, 54–5, 88
    data 82
    decommissioning 125
    high-efficiency lighting 50
    hydropower 18, 41
    natural-gas-fired power plants 15
    ocean energy 24
    photovoltaics 25, 27–8, 83–7
    renewable vs fossil fuels 2–3, 6–7
    solar water heating systems 33
    storage 41, 43–4
    technology 7–8, 13–14
    thermal insulation 47–8
    wind power 22
    see also investment costs; levelised cost of energy; marginal abatement cost; operation and maintenance costs
cost-benefit analysis (CBA) 2–3
    investor perspective 98–9
    social 205, 209, 230–8
cost-benefit ratio (CBR) 121–3, 230, 237–8
'cradle-to-grave' (life cycle) assessment 4, 211–21

data
    availability of 4, 79–82
    computer modelling 72–6, 79–83, 88, 90
    time-driven computer models 67

decarbonisation
    carbon capture and storage
        16–17
    energy storage 39
    policy 3–4, 203–5, 248
decision analysis. *see* multi-criteria
    decision analysis
decommissioning costs 125
decomposition (decision analysis)
    151
demand for energy 1, 233
    combined heat and power 34–5, 36,
        60–1, 65–6
    energy storage 39, 40, 42, 44, 60–1
    hydropower 17, 40
    modelling 62, *64*, 65–6, 88, *89*
    photovoltaics 84, *85*, *86*, *245*, 246,
        247
depreciation (financial
    analysis) 112–14, 120–1, 144
determinism 58, 69
discount factors (DF) 107–8, *108*,
    251–3
    net present value 124
    social cost-benefit analysis 235
discount rates (financial analysis) 107,
    109–12, *113*
    financial measures 117–8
    internal rate of return 127–31
    life cycle cost 131–3
    net present value 124–7
    social cost-benefit analysis 236
discounted payback periods 119–20
disjunctive method (decision
    analysis) 156
dispatchability 8, 210
dominance method (simple
    non-compensatory methods) 152,
    153–4, 201

EAC (equivalent annual cost) 114–15
economic appraisal techniques 2–5, 9.
    see also cost-benefit analysis;
    financial analysis
EDCs (engine-driven chillers) 39
Eigenvector method (decision
    analysis) 169, 173–5
ELECTRE TRI sorting method 188–9,
    195–8, 201

electrical power generation 14–28
    coal-fired power plants 15–17
    hydropower 17–19
    natural-gas-fired power plants
        14–15, *16*
    ocean energy 22–5, *24*, *25*
    photovoltaics 25–8, *26*, *28*
    wind power 19–22, *20*, *23*
electrical storage 40
emissions. *see* greenhouse gas emissions
emissions mitigation. *see* decarbonisation
energy demand. *see* demand, for energy
energy efficiency
    definition 12
    and technology 45–50
energy-efficient projects 1–2, 99–100.
    see also appraisal techniques
energy market 3, 7–8, 204–5, 233–4.
    see also financial analysis
energy systems, modelling 54–8,
    60–1
energy uses. *see* uses of energy
engine-driven chillers (EDCs) 39
engineering approach (financial
    analysis) 6, 59
environmental impacts of energy
    production
    cost-benefit analysis 98–9
    history 3–4
    life cycle assessment 211–21
    policy 203–5
    see also multi-criteria decision
        analysis
epistemic uncertainty 134
equity 99, 111, 206, 248
equivalent annual cost (EAC) 114–15
error (computer modelling) 72–6
European Union (EU) 207
event-driven computer models 67
Excel (*software*) 72, 83, 84, 89, 92
externalities (costs and benefits) 99, 230

feed-in-tariffs (FITs) 207, 225–9, 230,
    243–7
financial analysis (investment
    appraisal) 2–3, 6, 97–8, 148–9
    case studies 139–48
    and cash flow 97–8, 100–6
    discount rates 107, 109–12, *113*

financial analysis (investment appraisal)
(*cont'd*)
  financial measures 97–8, 116–17,
    136, *137–8*
  internal rate of return 127–31
  investor perspective 98–9
  levelised cost of energy 132–4,
    147–8
  life cycle cost 6, 131–2, 222–3,
    228–9, 243
  net present value 117, 123–7, *127,*
    129
  payback periods 7, 117–20
  photovoltaics 27–8
  present value 106–9
  profitability index and
    savings-to-investment
    ratio 121–3
  project types 99–100, 114–6
  real and nominal prices 104–6,
    109–10, 117–8
  return on investment 120–1
  taxation and depreciation 102,
    112–4, 120–1, 144
  uncertainty and risk 7, 109, 112,
    134–6, 238
  wind power 21
  see also modelling (computer)
financial appraisal techniques 2–5, 9.
  see also cost-benefit analysis;
    financial analysis
financial measures 97–8, 116–7, 136,
  137–8. see also individual named
  measures
financial ratios 116–7
  profitability index 121–3
  return on investment 120–1
  savings-to-investment ratio 122–3
FITs (feed-in-tariffs) 207, 225–9, 230,
  243–7
flow diagrams (computer modelling)
  62, *63, 66, 66*
fossil fuels 2–3, 5–7, 57, 82. *see also*
  individual named fuels

gas heat pumps (GHPs) 87–8, *88–90*
gas turbine plants 15, 232–3
gas-fired power plants 14–5, 58, 204,
  220–3

gasification 16, *17*
generation of energy
  CHP. *see* combined heat and power
  electrical. *see* electrical power
    generation
  heat. *see* heat generation
government policy. *see* policy
Great Recession of 2009 7
green taxes 7, 204–7, 223, 233
greenhouse gas (GHG) emissions
  carbon capture and storage 16–17
  energy storage 39
  life cycle assessment 211–21
  policy 3–4, 203–5, 207, 238–43, 248

Harmonised Index of Consumer Prices
  (HICP) 105, *106*
heat generation 28–34
  boilers 28–30, *29*
  solar water heating systems 12–3,
    30–1, 30–4, *32, 33, 34*
  storage of energy 44–5
heat of combustion 12
heat rate (HR) 15, *92*
heating value 12
hierarchies, analytic hierarchy
  process 168–81
high-efficiency lighting 48–50, *49*
higher (gross) heating value (HHV)
  12, 29, 35, *38*
hurdle rate 111
hybrid systems 57–8
hydroelectric storage 40–2
hydropower 17–9, 41

incremental cash flows 100–1
independent projects 99–100
index, profitability (PI) 121–3
inflation 104–5, *106,* 108
inflation indexes 105, *106*
input, energy 12
  computer modelling 55, 60, *85,* 88,
    92–3
input–output (I–O) analysis (life cycle
  assessment) 213–7, 220–1
instruments, policy 206–10
insulation, thermal 46–8, *47, 48*
  computer modelling 70, *70–1*
integrated gasification 16, *17*

interdependent projects 99–100
internal rate of return (IRR) 127–31
International Panel on Climate Change
    (IPCC) 203
international policy. *see* policy
investment appraisal. *see* financial
    analysis
investment costs 7–8
    cash flows 101–2
    coal-fired power plants 16
    compressed air energy storage 43
    hydropower 18–9
    natural-gas-fired power plants 15
    technology learning 13
investor perspective (financial
    analysis) 98–9
IRR (internal rate of return) 127–31

judgement matrices, analytic hierarchy
    process 176–81

Kaldor–Hicks criterion 230, 238
Kyoto agreement 3–4

latent heat 12
latent heat storage 45
laws. *see* policy
'learning-by-doing' (technology). *see*
    technology learning
least common multiple method 114,
    116
levelised cost of energy (LCOE) 132–4,
    147–8
    policy 209, 221–2, 226–9
lexicographic methods (decision
    analysis) 157–8, 201
LHV (lower (net) heating value) 12
life cycle assessment (LCA) 4, 211–21
life cycle cost (LCC) 6, 131–2, 222–3,
    228–9, 243
life cycle emissions (LCEs) 213,
    217–20, 222–3
life cycle inventory analysis 213–6
lifespan of project 97
    case study 103–4
    depreciation 112–4
    levelised cost of energy 132
    net present value 126
    unequal lifespans 114–6

lifespan, energy generators/
    technology
    boilers 28
    high-efficiency lighting 50
    hydroelectric storage 41
    hydropower 18–9
    natural-gas-fired power plants 15
    ocean energy 24
    solar water heating systems 33
    wind power 21
lighting, high-efficiency 48–50, *49*
load factor 13
lower (net) heating value (LHV) 12

maintenance costs. *see* operation and
    maintenance costs
marginal abatement cost (MAC) 4
    policy 209, 210–1, 221–4, 238–43
    subsidies 7, 205
market, for energy 3, 7–8, 204–5,
    233–4. *see also* financial analysis
market value, environmental/ social
    goods 3–4, 7–8, 204
market-based policy instruments 206
MARR (minimum acceptable rate of
    return) 111–2, 129, *129*
mathematical modelling 58–9. *see also*
    modelling (computer)
Maximax method (decision
    analysis) 160
Maximin method (decision
    analysis) 159–60
MDCA. *see* multi-criteria decision
    analysis
measurement (data) 76
meteorological data 82
micro-CHP (combined heat and power)
    36, *36–7*
Microsoft Excel (*software*) 72, 83–4,
    89, 92
minimum acceptable rate of return
    (MARR) 111–2, 129, *129*
mitigation of carbon emissions. *see*
    decarbonisation
modelling (computer) 4, 9, 53–4,
    93–5, 97
    analysis 68–71, 79, 93
    case studies 83–93
    cash flows 59, 83–7, 90–3, 93–4, 97

modelling (computer) (*cont'd*)
   data sources 79–83
   energy projects 76–9
   models 58–71
   simulation 71–6, 79, *80–1*, 83–5,
      88–9, 92–3
   systems 54–8
modified internal rate of return
   (MIRR) 129–31
Monte Carlo simulation (risk) 135
multi-criteria decision analysis
   (MDCA) 3, 9, 151–2, 200–1
   analytic hierarchy process 168–81,
      201
   concordance analysis 181–9
   policy 3–4, 209–10, 248
   simple additive weighting
      method 160–8
   simple non-compensatory
      methods 152–60
   wind farm case study 189–200
mutually exclusive projects 99–100,
   126, 129

national policy. *see* policy
natural systems 57
natural-gas-fired power plants 14–5,
   58, 220–3
net operating cash flow 102
net present value (NPV) 117, 123–7
   cash flows 106–9, 123–7
   equivalent annual cost 115
   policy 209, 229–30, 236–7,
      245–6
   profile 127, *127, 129*
   risk 135–6
nominal vs. real prices 104–6, 109–10,
   117–8
non-compensatory methods (decision
   analysis). *see* simple
   non-compensatory methods
non-economic appraisal
   techniques 3–5, 9. *see also* life
   cycle assessment; multi-criteria
   decision analysis
NPV. *see* net present value

ocean energy 22–5, 58, 115–6
oil prices 2–3

operation and maintenance (O&M)
   costs
   boilers 29
   cash flow 102
   hydroelectric storage 41
   hydropower 18–9
   natural-gas-fired power plants 15
   photovoltaics 28
   solar water heating systems 33
   wind power 22, *66*
opportunity costs (cash flow) 101
optimisation (computer modelling)
   69–70, *70–1*
output, energy 12–13
   computer modelling 55, 60, *85*, 88,
      92
over and above costs (cash flow) 101

Pareto improvement 230, 237
payback periods 7
   financial analysis 117–20
   thermal insulation 47–8, *48*
penstocks 17–8, 41
personal computers (PCs) 4. *see also*
   modelling (computer)
phase change materials (PCM) 45
photovoltaics 25–8
   computer modelling 67–8, 83–7
   policy 243–7
PI (profitability index) 121–3
plant availability factor 13
policy 1, 7–8, 9, 203–5, 248
   case studies *206*, 238–47
   coal-fired power plants 16–7
   instruments and targets 206–10
   life cycle assessment 211–21
   marginal abatement cost 210–1,
      221–4
   multi-criteria decision analysis 3–4,
      209, 210, 248
   and project net present value 124
   social cost-benefit analysis 230–8
   subsidies 204–5, 207–8, 224–30
   wind farm case study 189–94
power generation
   CHP. *see* combined heat and power
   electrical. *see* electrical power
      generation
   heat. *see* heat generation

power plants
  availability factor 13
  capacity factor 13, 19, 131, 220
  coal-fired 15–7, 204
  combined cycle gas turbine 15,
    232–3
  gas-fired 14–5, 58, 204, 220–3
  hydropower 17–9
  ocean energy 22–5, 58, 115–6
  photovoltaics 25–8
  rated power 12–3
  see also combined heat and power;
    wind power
practitioner readership 5
prerequisite projects 100
present value 106–9, 209, 237, 246
  discount rates 110, 112, *113*, 131–2
  equivalent annual cost 115
  profitability index 122
  savings-to-investment ratio 122–3
  see also net present value
price of energy. *see* cost of energy
prioritisation, analytic hierarchy
  process 169–75
process analysis (life cycle
  assessment) 217–21
process-based hybrid analysis (life cycle
  assessment) 220–1
profitability index (PI) 121–3
programming languages 72
programs, computer modelling 72–3,
  78, *80–1*
project appraisal. *see* appraisal
  techniques
projects, energy-efficient 1–2, 99–100.
  see also appraisal techniques
PROMETHEE I decision
  method 184–8, 201
public policy. *see* policy
pumped hydroelectric storage
  (PHS) 17, 21, 40–2, 122

rated power 12–3
rates of return 116–7
  internal rate of return 127–31
  minimum acceptable rate of return
    111–2, 129, *129*
  modified internal rate of return 129–31
  and net present value 126–7

real vs. nominal prices 104–6, 109–10,
  117–8
recession of 2009 7
regulations/ standards 207. *see also*
  policy
replacement chain method (least
  common multiple method) 114,
  116
report (computer modelling) 79
representativeness (computer
  modelling) 72–6
residual values (financial analysis) 102
resistance-to-change grid (decision
  analysis) 167–8
return on capital employed
  (ROCE) 121
return on investment (RoI) 120–1
return rates. *see* rates of return
revealed preference techniques 234
revenues 6–7, 101–2
  compressed air energy storage case
    studies 43, 90–3, 124–6
  computer modelling 59, 60, *85–6*
  feed-in-tariffs 226, 244–6
  lifespan of project 103
  wind farm case studies 142, 144,
    145, 148, 229
risk (financial analysis) 7, 109, 112,
  134–6
ROCE (return on capital employed)
  121
RoI (return on investment) 120–1

satisficing methods (decision
  analysis) 155–6
savings-to-investment ratio
  (SIR) 122–3
scenario analysis (risk) 134–5
sector-specific data 82–3
sensible heat storage 44–5
sensitivity analysis (risk) 134–6
sensitivity analysis (simple additive
  weighting analysis) 163–4
sequential elimination methods
  (decision analysis) 157–8
shadow price factors (SPF) 234
signal (modelling) 62–3, *86, 88–9*, 91
simple additive weighting (SAW)
  method (decision analysis) 160–8

simple non-compensatory methods
(decision analysis) 152–60
attitude-oriented methods 158–60
dominance 153–4
satisficing methods 155–6
sequential elimination
methods 157–8
simulation 71–6, 79, *80–1*, 83–5,
88–9, 92–3. *see also* modelling
(computer)
SIR (savings-to-investment ratio) 122–3
SMP (system marginal price) 90–4
societal impacts of energy production
history 3–4
investor perspective 98–9
social cost-benefit analysis 205, 209,
230–8
see also multi-criteria decision
analysis
software, computer modelling 72–3,
78, *80–1*
solar water heating systems
(SWHS) 12–3, 30–4
standards/ regulations 207. *see also*
policy
stated preference techniques 234
statistics, computer modelling 73–6
steam turbine generation system 16
stochasticism 58
storage of energy 8, 39–45
compressed air energy 42–3, *43*
computer modelling 55, 60–1
electrical 40–3
hydroelectric 17, 21, 40–2
latent heat 45
sensible heat 44–5
thermal 44
straight-line depreciation 113–4
student readership 5
subsidies 7–8
cash flow 102
marginal abatement cost 7
policy 204–5, 207–8, 224–30
substitution (cash flow) 101
sunk costs (cash flow) 100–1
supply of energy 5–6
cash flow 102
energy storage 39
see also use of energy

system marginal price (SMP) 90–4
systems (computer modelling) 54–8

targets, policy 206–10
tariffs, feed-in (FITs) 207, 225–9, 230,
243–7
tax on carbon 7, 204–7, 223, 233
taxation (financial analysis) 102, 112–4
return on investment 120–1
technology 8–9, 11–4
combined heat and power 34–9
costs 7–8
electrical power generation 14–28
energy efficiency 45–50
heat generation 28–34
marginal abatement costs 221–4
policy 204–5, 221–4
storage of energy 39–45
technology learning 13–4
photovoltaics 25
policy 224
wind power 22
temporal measures 116–7
thermal efficiency 15
thermal insulation 46–8, *47*, *48*
computer modelling 70, *70–1*
thermal storage 44, *46*
tidal turbines *24*
time value (cash flow) 109, 121
time-driven computer models 67
transformation (modelling) 62–5, *63*,
86, *88–9*, 91, *92*
turbines, CHP (combined heat and
power) 37
turbines, tidal *24*
turbines, wind *20*, 20–1
computer modelling 65–6

uncertainty (financial analysis) 134–6,
238
use of energy 13–4
coal-fired power plants *17*
data sources 79–82
gas-fired power plants *16*
high-efficiency lighting 50
hydroelectric storage 41–2
hydropower 17, 19, *20*
ocean energy 24–5
photovoltaics 25–6, *26*

solar water heating systems *33*, 33
wind power 19, 22, *23*
see also supply of energy; demand for
    energy

validation (computer modelling)
    72–3, 76, *77*, 79
case studies 84–5, 89–90, 92–3
verification (computer modelling)
    72–3, 76, *77*
case studies 83–5, 89–90, 92–3

water power (hydropower) 17–9, 41
wave energy (ocean energy) 22–5, 58,
    115–6
weighted average cost of capital
    (WACC) 110–2
case studies 144–5, *147*, 148

discount rate 125
internal rate of return 129–30
net present value 127, 148
sensitivity analysis 135
weighting (simple additive weighting
    analysis) 164–7
welfare impacts. *see* societal impacts of
    energy production
willingness to accept (WTA) 234
willingness to pay (WTP) 234, *235*
wind power 19–22, *20*
capacity of 19, 22, *23*
case studies 142–8, 189–200, 232
computer modelling 65–6, 74–6
depreciation 112–3
return on investment 120, *121*
wind turbines *20*, 20–1
computer modelling 65–6